AF573772

Alzheimer's Disease – Modernizing Concept, Biological Diagnosis and Therapy

Advances in Biological Psychiatry

Vol. 28

Alzheimer's Disease – Modernizing Concept, Biological Diagnosis and Therapy

Volume Editors

H. Hampel Frankfurt a.M.

M.C. Carrillo Chicago, Ill.

9 figures, 1 in color, and 14 tables, 2012

KARGER

Basel · Freiburg · Paris · London · New York · New Delhi · Bangkok · Beijing · Tokyo · Kuala Lumpur · Singapore · Sydney

Advances in Biological Psychiatry

Prof. Harald Hampel, MD, MA, MSc
Goethe University
DE–60528 Frankfurt a.M. (Germany)

Maria C. Carrillo, PhD
Alzheimer's Association
Chicago, IL 60601 (USA)

Library of Congress Cataloging-in-Publication Data

Alzheimer's disease : modernizing concept, biological diagnosis and therapy / volume editors, H. Hampel, M.C. Carrillo.
p. ; cm. -- (Advances in biological psychiatry, ISSN 0378-7354 ; v. 28)
Includes bibliographical references and index.
ISBN 978-3-8055-9802-6 (hard cover : alk. paper) -- ISBN 978-3-8055-9803-3 (electronic version)
I. Hampel, H. (Harald), 1962- II. Carrillo, M.C. (Maria C.) III. Series: Advances in biological psychiatry ; v. 28. 0378-7354
[DNLM: 1. Alzheimer Disease--diagnosis. 2. Biological Markers--metabolism. W1 AD44 v.28 2012 / WT 155]

616.8'31--dc23

2012002099

Bibliographic Indices. This publication is listed in bibliographic services, including Current Contents® and Index Medicus.

www.karger.com
Printed in Germany on acid-free and non-aging paper (ISO 9706) by Bosch Druck, Ergolding
ISSN 0378–7354
e-ISSN 1662–2774
ISBN 978–3–8055–9802–6
e-ISBN 978–3–8055–9803–3

Contents

Hampel H, Carrillo MC (eds): Alzheimer's Disease – Modernizing Concept, Biological Diagnosis and Therapy.
Adv Biol Psychiatry. Basel, Karger, 2012, vol 28, pp 1–14

The Global Impact of Alzheimer's Disease

Maria C. Carrillo[a] · William Thies[a] · Lisa J. Bain[b]

[a]Alzheimer's Association, Chicago, Ill. and [b]Independent Science Writer, Elverson, Pa., USA

Abstract

Alzheimer's disease is a worldwide epidemic, however the causes remain elusive. This chapter provides an overview of the numbers of individuals affected by Alzheimer's disease and the impact it has on the family. It also explores the current thinking on the causes and genetics that give researchers clues to possible predictors and treatment options, both current and investigational. The most current framework of Alzheimer's disease in which silent biological changes can begin perhaps decades before symptoms is explored. Additionally, potential biomarkers that could detect the biological changes before symptoms appear are described.

Dementia was estimated to affect 35.6 million people worldwide in 2010, and if nothing is done to prevent or slow the disease, the prevalence of dementia is expected to increase to 65.7 million by 2030 and 115.4 million by 2050 [1]. The impact of this exponential growth in the number of people with dementia will be felt around the world, particularly in low- and middle-income countries where nearly two-thirds of affected people live and where the increases are expected to be most profound.

The main reason for this worldwide explosion of dementia prevalence is the aging of the population. In 2006, about 8% of the world's population, or 500 million people, were over age 65; yet by 2030, those over age 65 are expected to comprise 13% of the total population. Developing countries are expected to see the most rapid and dramatic expansion of older populations, with increases of as much as 140% anticipated compared to an increase of only about 51% in developed countries [2]. And the fastest growing segment of the population is expected to be the oldest-old, those 85 years of age and older; this population group is expected to increase 151% worldwide between 2005 and 2030 [3]. These statistics reflect the fact that people around the world are living longer and healthier lives as a result of medical advances that have reduced the incidence of communicable diseases. At the same time, chronic, non-communicable age-related diseases including Alzheimer's disease (AD) and other dementias are on the rise.

AD is the most common type of age-related dementia, accounting for approximately 60–70% of all dementias [4, 5]. The disorder that would eventually bear his name was first described by a German neurologist, Dr. Alois Alzheimer, in 1906 [6]. Following the death of a patient in her early 50s who had suffered from memory loss and difficulty both speaking and understanding what was said to her, Dr. Alzheimer performed a detailed study of her brain, noting dramatic brain atrophy as well as unusual protein deposits called amyloid plaques and neurofibrillary tangles. More than a century later, these neuropathological characteristics are still recognized as the hallmarks of AD. We now know that AD is progressive, incurable, and fatal, caused by the degeneration of cells in distinct brain regions and that the disease process begins as many as 10–20 years before outward symptoms become apparent. Memory loss is usually the first symptom of AD, but over time a person with the disease experiences other gradually worsening symptoms, including changes in personality, a decline in the ability to speak and understand what others say, poor decision-making ability and an inability to care for oneself.

Age is the most important risk factor for AD, with studies indicating that throughout the world, the incidence of dementia doubles approximately every 5–6 years between the ages of 65 and 90 [7], and continues to increase exponentially even beyond age 90, reaching as high as 41% per year among those over age 100 [8]. The dramatic increase in size of the aging population in developing countries combined with the elevated risk of dementia associated with advancing age results in some sobering predictions. In their World Alzheimer's Report 2010, Alzheimer's Disease International (ADI) forecasted that by 2050, 71% of people with dementia worldwide would live in low- and middle-income countries, compared to 58% today. Europe is expected to see a 40% increase in the number of people with dementia, and North America about a 63% increase. Other developed countries can expect similar increases. It is in the developing world where the increases are expected to be most dramatic: 107 and 117% in South and East Asia, respectively; 125% in North Africa and the Middle East, and 134–146% in parts of Latin America [9].

Shouldering the Burden of Dementia Care

The astronomical numbers of people with dementia will put a tremendous strain on governments and public health systems around the world. In 2010, total estimated worldwide costs of dementia were USD 604 billion, about 1% of the world's gross domestic product. This figure exceeds the annual revenue of the world's largest companies, Walmart (USD 414 billion) and Exxon Mobil (USD 311 billion). If it were a country in and of itself, dementia care would rank 18th in the size of its economy, nearly the same as the economy of Turkey [9].

The cost of dementia care includes direct costs of medical care, direct costs of social care, and informal costs. About 70% of the costs of caring for people with dementia occur in Western Europe and North America where direct costs of social care predominate. This includes costs of resource- and cost-intensive residential and nursing home care and professional care in the community. In contrast, informal care costs predominate in lower income countries, where care tends to be provided by family members and others. The per person cost of caring for a person with dementia is similarly disproportional: nearly USD 50,000 per person in North America compared to less than USD 1,000 in South Asia and Western Sub-Saharan Africa [9].

As the prevalence of dementia increases worldwide, the costs of caring for people with dementia will likewise soar. In just 4 years between 2005 and 2009, the worldwide societal costs of dementia ballooned by an estimated 34% (18% in fixed costs) [10], and a tentative estimation by ADI predicted an 85% increase in costs by 2030, with costs in developing countries rising faster than in high-income countries [9].

The burden of dementia care falls heavily on families and caregivers, particularly in low- and middle-income countries where people with dementia tend to live at home rather than in long-term residential or nursing home care facilities. Unfortunately, there has been little research on caregiver burden, resource utilization, and awareness in the developing world. Recognizing that less than one-tenth of all population-based research has been directed to the two-thirds of people with dementia who live in developing countries, the 10/66 Dementia Research Group was established in 1998 to redress this imbalance [11]. Studies conducted by the Group have identified numerous areas of concern as developing countries struggle to cope with the rapidly growing problem of dementia: a low level of awareness about the disease; lack of appropriate, affordable, and culturally acceptable healthcare services, and a heavy reliance on family members who themselves may not have the support or resources to manage these responsibilities [12].

From a policy perspective, worldwide responses to this impending disaster vary considerably as well. For example, in India, the parliament passed a law in 2007 that required children to support their parents or face prison [9]. As a byproduct of economic development and increased personal income, care homes are proliferating in many developing countries, including India and China, although these tend to cater primarily to the affluent. In Japan where there is a strong tradition of honoring the elderly; the government implemented a long-term care insurance plan that entitles individuals to services [13]. Despite initial misgivings about the cultural appropriateness of this plan, it has proven popular and appears to have minimized caregiver strain [14].

Bending the Curve

Around the world, public health experts and policymakers are scrambling to develop strategies to address this global public health crisis without bankrupting governments already stretched to the limits by the current global financial crisis. The most effective strategy, indeed perhaps the only approach that can halt this inexorable epidemic that threatens global stability is to find a way to cure or slow the disease. Brookmeyer et al. [15] estimated in 1998 that if the onset of AD could be delayed by 5 years, the prevalence of the disease would decrease by almost 50% by the middle of the century. Even delaying onset by as little as 1 year would have enormous public health implications, reducing the worldwide prevalence by nearly 9.2 million cases by 2050, with most of the decline among those who need the highest level of care [16].

Recent progress in understanding the fundamental molecular mechanisms underlying aging, dementia, and AD suggests avenues to pursue towards delaying onset or even curing the disease [17]. Given that it is now widely accepted that the pathological process begins long before symptoms appear, the first step along this pathway requires early diagnosis, leading to early intervention. On the treatment front, multiple pathways have been implicated in the disease, suggesting new therapeutic targets. And finally, epidemiological research has identified modifiable vascular and lifestyle risk factors as well as protective factors that suggest prevention may be a realistic goal.

New Criteria for Diagnosing Alzheimer's Disease

The current criteria for diagnosing AD were established in 1984 [18] when little was known about the genetic and molecular events that result in disease, and before imaging technologies that could reveal the dysfunction or death of brain cells had become available. Developed by the National Institute of Neurological and Communicative Disorders and Stroke (NINCDS, now NINDS) and the Alzheimer's Disease and Related Disorders Association (ADRDA, now the Alzheimer's Association), these criteria required a physical evaluation as well as neuropsychological testing to classify patients as having *probable* or *possible* AD. Although a definite diagnosis was made only after confirmation by a detailed study of the brain tissue at autopsy, a diagnosis of probable AD required progressive impairment in at least two of eight cognitive domains. Memory is the most common of these domains to be affected; others include concentration, problem-solving and language.

Now, thanks to the studies such as the Alzheimer's Disease Neuroimaging Initiative (ADNI), the Australian Imaging, Biomarker, and Lifestyle Flagship Study of Ageing (AIBL) and global efforts known as World Wide ADNI, biomarkers have been identified that offer the potential for making not only an earlier diagnosis but one that provides more precise information about the pathological basis of the disease and thus, may help identify appropriate interventions. As a result of this improved understanding of the disease, the National Institute on Aging (NIA) and the Alzheimer's Association convened three workgroups to update the diagnostic criteria so that they would better reflect the full range of the disease from its earliest effects to its eventual impact on mental and physical function. The new criteria and guidelines were first presented at the Alzheimer's Association International Conference on Alzheimer's Disease in 2010, and after soliciting comments from the Alzheimer's community, the proposed criteria and guidelines were published in April, 2011 [19–22].

The proposed criteria and guidelines identify three different phases of the disease: (1) A pre-symptomatic, or pre-clinical, phase occurring many years before symptoms become evident. (2) A symptomatic phase characterized by mild problems in cognition, learning, and memory, enough to be noticed and measured but not enough to impair the ability to live independently or carry out everyday activities. This phase is what is currently referred to as mild cognitive impairment (MCI) and includes four levels of certainty, depending on biomarker findings. Thus, individuals with very mild cognitive changes but positive biomarker findings but nonetheless receive a diagnosis of MCI. (3) Probable and possible Alzheimer dementia, when the disease has progressed to the point that the person has markedly impaired memory and cognition and cannot function independently.

Importantly, the criteria for the MCI and dementia classifications were designed to serve both general healthcare providers who may not have access to advanced biomarker technologies or detailed neuropsychological testing, as well as investigators in specialized AD centers, while the criteria for preclinical AD are intended purely for research purposes. Indeed, it is during this long preclinical phase that biomarkers provide important evidence about early molecular events that may lead to AD pathology and thus may suggest new therapeutic targets.

Role of Biomarkers in Diagnosing and Treating Alzheimer's Disease

Since Alois Alzheimer identified amyloid plaques and neurofibrillary tangles in the brain of his patient, AD has been defined as an amyloid-based disease. In 1991, John Hardy and David Allsop [23] first proposed the amyloid cascade hypothesis, which proposed that deposition of beta-amyloid (Aβ) and tau phosphorylation were key events that lead to neuronal death in AD. New biomarker technologies now allow assessment of Aβ deposition and neuronal injury non-invasively through positron emission tomography (PET) with amyloid-binding ligands and by measuring the concentration of Aβ and phosphorylated tau (p-tau) and total tau in the cerebrospinal fluid (CSF) [24]. A recent study from ADNI identified an 'Alzheimer's disease signature' based on measurements of Aβ and p-tau in the spinal fluid, which appeared to predict whether a person with MCI would go on to develop AD within 5 years [25]. Interestingly, the signature was also present in more than a third of cognitively normal subjects, suggesting that AD-related changes in the brain may be detectable even before problems with memory and thinking appear.

Other biomarkers assess synaptic dysfunction using FDG-PET imaging, regional brain degeneration using structural CT and MRI, changes in regional brain activity assessed by functional neuroimaging techniques, regional changes in brain perfusion, changes in various gene markers, neuropsychological markers of early brain dysfunction, and a host of other biochemical changes in the CSF. For example, a recent study used a target proteomic screen to search for novel candidate biomarkers, identifying 37 analytes that are altered in the CSF of individuals with early AD, and one (calbindin), that appeared to predict the risk of future dementia in cognitively normal participants [26].

Treating Alzheimer's Disease

Antecedent biomarkers are needed not only to identify people before dementia symptoms arise, but also to identify subjects for specific types of clinical studies and to monitor a subject's response to treatment. No treatment has thus far been shown to slow the progression of AD, although there are several currently available drugs that appear to improve cognition and memory, and other drugs that can relieve some of the more troubling symptoms, such as depression and agitation.

The most widely used drugs for treating AD are the cholinesterase inhibitors, which prevent the breakdown of acetylcholine, a neurotransmitter that plays an important role in memory and other mental functions but that is depleted in AD when the cells that produce it are destroyed. Cholinesterase inhibitors that are approved by the United States Food and Drug Administration (FDA) include Aricept® (donepezil), Razadyne® (galantamine), and Exelon® (rivastigmine). Drugs that inhibit the breakdown of acetylcholine can only compensate for lowered levels of the chemical in the early stages of the disease before there has been massive loss of the cells that produce the chemical. Randomized clinical trials of these drugs have produced small to moderate evidence of benefit in a poorly characterized subgroup of patients, with clear evidence of adverse effects [27].

Another FDA-approved therapy for AD is Namenda® (memantine). Memantine works by interfering with NMDA (N-Methyl-D aspartate) receptors, resulting in decreased levels of glutamate, which can destroy brain cells. When given to people with moderate-to-severe AD, this drug can allow them to function more independently for several months longer than they might without the drug. However, a recent study concluded that memantine was ineffective in mild AD and produced only small improvements in cognition in patients with moderate AD [28].

Despite the fact that cholinesterase inhibitors are only FDA approved for mild AD and memantine for moderate-to-severe AD, clinicians frequently prescribe these medications off-label for patients in various stages of the disease. A recent observational study, however, suggests that these drugs may hasten cognitive decline in AD and MCI patients, and that combination therapy with both drugs may be even worse [29].

These results clearly show that new approaches to treating AD are needed. AD is a complex illness, with multiple promising avenues for therapeutic intervention.

Indeed, it is unlikely that targeting a single pathway will result in a broadly efficacious therapy or cure. The hunt for new therapies is intimately tied to better understanding the fundamental mechanisms underlying the disease and to the discovery and validation of novel biomarkers of disease progression. In terms of mechanism, the amyloid cascade hypothesis has dominated the drug development and discovery process for many years [23], with many drugs developed to prevent the production of Aβ, interfere with its formation into plaques, or increase the rate of its degradation and clearance from the brain. Unfortunately, although these efforts have given rise to a number of drugs that have gone through several stages of clinical trials, none has yet proven to be effective.

The amyloid cascade hypothesis also proposes that Aβ aggregation triggers a number of downstream events, including phosphorylation of the protein tau, and this leads to the formation of neurofibrillary tangles and the death of neurons. Thus, another avenue of drug development has produced strategies that prevent the chemical modification of tau to its toxic form (p-tau), the formation of p-tau tangles, or binding of p-tau so it is not able to exert its toxic effects. Again, so far, none of these drugs have yet been shown effective. The reason for these failures may be the drugs were given too late in the disease process. However, another possible reason is that molecular processes other than those involving amyloid may contribute to AD [30].

New Disease Concepts

There are many pathways involved in nerve cell death or dysfunction. Among the mechanisms that have been implicated in AD are calcium dysregulation, proteolysis failure, altered cell signaling, mitochondrial dysfunction and oxidative stress, and inflammation, all of which can lead to the central features of AD – synaptic dysfunction and neurodegeneration [30].

Inflammation is a cardinal feature of the neuropathology in the AD brain [31], suggesting that the inflammatory process could play a role in neurodegeneration. Mitochondria have also been implicated as being central to the AD process [32]. Mitochondria play important roles in many cellular processes including apoptosis and synaptogenesis, and also produce reactive oxygen species that damage cells through oxidative stress. Mitochondrial abnormalities have been seen in the brains of people with AD, and some research suggests that targeting mitochondrial dysfunction and oxidative stress could slow or prevent the nerve damage seen in AD.

One approach to elucidating the mechanisms of disease, particularly one as common and complex as AD, is to examine risk factors As mentioned earlier, age is the strongest risk factor for dementia, but many other conditions also substantially elevate AD risk. Understanding these risk factors may thus be the first step toward developing new treatments and/or preventive strategies.

Genetics

Late-onset AD (LOAD), with onset of symptoms after age 65, is the most common form of AD. It is not considered a genetic disease because no single gene determines if an individual will develop the illness. However, genetics undoubtedly play a role in the disease because having a first-degree relative with AD increases the risk of getting the disease. In addition, a rare form of the disease called early-onset familial AD (eFAD), which typically appears before age 50 (sometimes as early as the late 20s), is hereditary – if a parent has the disease, his or her offspring have a 50–50 chance of also being affected. Most cases of eFAD have been linked to mutations in presenilin-1 (PS1), but at least two other genes, PS2 and amyloid precursor protein (APP), have also been linked to eFAD. Genetic testing can be used to predict whether a person will develop eFAD; however, many people choose not to get genetic testing for a progressive, incurable and ultimately fatal disease.

Other genes also increase the risk of developing LOAD, particularly the *APOEε4* gene variant, which is the strongest genetic predictor of risk. APOE shuttles cholesterol, other fat-carrying proteins and other fat-soluble substances through the bloodstream. A person's risk of developing LOAD increase about 12- to 15-fold if he or she has two copies of the *APOE-ε4* gene, and these people also have a much younger age of onset than those people who develop AD but have only one copy or no copies of this variant [33].

Other genes that elevate the risk of developing LOAD are *SORL1*, which is involved in transporting amyloid protein, and *TOMM40*. *TOMM40* resides near the *APOE* gene and may have an even stronger relationship with age of onset than *APOE* variants do. *TOMM40* exists as either a short, long or very long form, with long forms associated with earlier age of onset. Even more recently, variants in a gene called *FTO*, which is associated with obesity and diabetes, were shown to be associated with an increased risk of developing AD, especially in the presence of *APOEε4* [34].

Genetics may also play a protective role in the development of AD. Relatives of non-demented elderly people face a risk 11 times lower than the risk for relatives of people with AD [35]. Understanding these protective genetic factors may also offer clues to disease mechanisms.

Diabetes/Insulin Resistance

As with AD, the incidence of obesity, diabetes and insulin resistance is skyrocketing around the world. Many studies have shown that all of these metabolic conditions increase a person's risk of developing AD or cognitive impairment. In one study, diabetes was shown to increase the risk of developing AD by 65% [36]. However, in the Framingham Study, diabetes was not shown to increase the risk of developing AD [37].

Insulin resistance refers to an inability of the cells in the body to dispose of blood glucose through the action of insulin. This causes the pancreas to secrete more insulin, leading to hyperinsulinemia. When the pancreas cannot keep up with the need for more insulin, a person can develop glucose intolerance, or an inability to use glucose efficiently as a body fuel. Glucose intolerance is also called 'pre-diabetes', and if untreated can lead to diabetes. Obesity, or adiposity, is associated with a higher risk of insulin resistance, and there is evidence that weight loss can lower insulin resistance. This is important with regard to AD because insulin resistance and high insulin levels are associated with a higher risk of cognitive decline, particularly problems with memory. Some studies suggest that this association is even stronger in people with the *APOEε4* gene variant. Insulin resistance also contributes to cardiovascular disease, which also increases the risk of AD.

One possible link between insulin resistance and AD could be a chemical in the body called insulin-degrading enzyme, which has been shown to degrade not only insulin, but Aβ as well. Decreased function of this enzyme could theoretically increase the risk of AD, although this has not been proven [38]. Complicating the situation, while insulin has been shown to affect Aβ metabolism, there is evidence that diabetes is not associated with an increased amount of plaques and tangles [39]. Nonetheless, insulin and insulin-degrading enzyme may affect cognition and the development of AD independently of plaque formation. It may be that insulin resistance is related to increased oxidative stress, and this may be the link to cognitive dysfunction and AD. Other contributing factors may be substances produced by fatty tissue that are important in metabolism and inflammation, processes that have also been linked to brain health and cognition.

Cardiovascular Disease

Obesity is a risk factor not only for diabetes, but for cardiovascular disease as well. In addition, other vascular risk factors such as hypertension, high cholesterol and atherosclerosis have been associated with both vascular dementia and Alzheimer dementia. Indeed, the link between the presence of *APOEε4* and AD suggests a possible vascular-related mechanism for AD, and coronary artery disease has been associated with plaques and tangles in brain tissue, particularly in *APOE-ε4* carriers [40].

Brain Injury

Traumatic brain injury (TBI) has also been linked to an increased risk of dementia, particularly among athletes and soldiers, although there appear to be neuropathological differences between AD and dementia resulting from chronic traumatic encephalopathy, with more tau pathology among the latter [41]. In one study, 30% of people (some of them children) who died as a result of a TBI had Aβ plaques in their brains, suggesting that deposition of Aβ is an acute response to brain injury [42]. A review of the literature on TBI and neurodegenerative disease revealed that AD is associated with moderate and severe TBI, but not mild TBI unless there was a loss of consciousness [43]. One study of retired professional football players found that those players who had sustained three or more concussions were 5 times more likely to develop MCI and 3 times more likely to report significant memory problems. People with recurrent concussions were also associated with an earlier onset of AD symptoms compared to the general (male) population [44].

Depression

Depression has been linked to an increased risk of AD. According to researchers, people with a history of depression are 2.5 times more likely to develop AD than people who have never had depression, and this association appears to be independent of the incidence of vascular risk factors and stroke. Those who first develop depression before age 60 were especially vulnerable to AD, with a nearly 4-fold increased risk [45].

Estrogen

A link between declining estrogen levels and an increased risk of AD has long been suspected. Estrogen is known to have positive effects on learning, memory and mood, and also influences development and degeneration in the brain. It is also thought to be neuroprotective, affecting several pathways that influence the survival of neurons, including energy production, inflammation and IGF-1 (insulin growth factor) signaling [46]. IGF-1 levels are increased in LOAD, and it is thought that this is important in disease development, possibly a sign that production is increased because brain cells are resistant to its effects [47].

Because estrogen affects brain areas that are important for cognition, and because menopause is associated with sharp declines in the concentration of circulating estrogen, it was hypothesized that a decline in estrogen after menopause could be linked to cognitive decline and an increased risk of AD. And indeed, early studies suggested that estrogen replacement therapy could protect against cognitive aging in women [48]. However, the Women's Health Initiative Memory Study (WHIMS) showed that estrogen plus progestin therapy had an adverse rather than protective effect on cognition among women over 65 years of age, particularly in women with lower cognitive function at the beginning of the trial [49]. This failure of estrogen therapy to improve cognition among older women led to a new idea called the 'critical period hypothesis', which holds that estrogen therapy is neuroprotective, but only if initiated during a critical time period, close to when natural or surgically-induced menopause begins [50]. Many studies have supported this hypothesis [51], and it is currently being investigated in at least two clinical trials to determine whether estrogen therapy given in midlife, shortly after menopause begins, may reduce the risk of developing dementia [52].

As with other studies of risk factors, knowing the mechanism behind increased or decreased risk is essential for the development of therapies. In the case of estrogen, its beneficial effects on brain aging and cognition appear to be related to interactions with cholinergic neurons, and studies suggest that the response of these cholinergic neurons to estrogen declines with age, which could explain why estrogen does not benefit cognition in elderly women.

Is Prevention a Realistic Goal?

Many people believe that prevention is the only way to stop the disease from overwhelming our public health system and national economy. The path to prevention will involve a number of strategies including treating other conditions such as those mentioned above that increase the risk of dementia as well as developing new treatments – based on a better understanding of the molecular events that take place early in the neurodegenerative process – that will halt the disease in its earliest stages. In addition, there is substantial evidence that lifestyle modifications may be beneficial.

For example, the Nun Study is a long-term study of aging that began in 1986, enrolling 678 members of the School Sisters of Notre Dame between the ages of 75 and 102 [53]. In addition to taking part in numerous physical and neuropsychological evaluations, participants agreed to donate their brains for neuropathological studies upon death. These studies confirmed what had been seen in earlier studies, that the brains of some people who appear cognitively normal are, in fact, riddled with the same plaques and tangles seen in those with severe Alzheimer dementia.

Individual differences in the brain's capacity to cope with neurodegeneration, or cognitive reserve, has been proposed as the reason that some individuals are able to avert cognitive loss in the presence of significant neuronal loss [54]. Importantly, life experiences such as education, occupation, social interactions and leisure activities appear to increase a person's cognitive reserve, such that he or she is better able to tolerate brain damage or brain disease. In the Nun Study, for example, researchers demonstrated a correlation between linguistic ability (i.e. the expression of complex ideas) in early life and protection against cognitive problems in late life, even in the presence of disease [55]. Other studies have demonstrated that the risk of AD is associated with other lifestyle factors that are potentially modifiable, including education, mentally demanding occupations, cognitive and leisure activities, exercise, and diet [56]. A systematic review of the literature found that dementia risk was decreased by 46% in persons with high reserve, with complex mental activities having the greatest effect [57].

The introduction of the NIA-Alzheimer's Association revised criteria and guidelines provides a framework that offers a potential platform to identify people that will develop AD later in life, during an asymptomatic phase of the disease [19, 21, 22]. This early identification of individuals either at risk or already experiencing biological/pathological markers of AD without symptoms introduces the possibility of early intervention during this asymptomatic phase [19]. The guidelines discussed for a preclinical stage of AD are recommendations for the research community and are yet to be further investigated and validated. However, the proposed framework provides a hopeful stage for future therapeutic trials in asymptomatic populations.

Conclusion

AD represents a worldwide public health crisis of unimaginable proportions, demanding a massive, integrated, multidisciplinary, and sustained worldwide response. Over the past decades, there has been significant progress in understanding the molecular and biochemical basis of the disease and in applying that knowledge to the development of new technologies and treatments. The failure of these advances to slow the prevalence of the disease indicates that increased attention from governments, academic institutions, and the private sector, combined with a broadened perspective, are needed. Only through such research will better treatments – and eventually a cure – be found. Even in the face of the worldwide financial crisis, governments must take a leadership role in investing in Alzheimer research at a level that matches its impact on our economy and society – it must demonstrate the commitment to fighting AD through research, education and care.

References

1 Alzheimer's Disease International: World Alzheimer's Report, 2009.
2 National Institute on Aging: Why Population Aging Matters: a Global Perspective. NIH Publ. No. 07-6134. Bethesda, National Institute on Aging, 2007.
3 United Nations: World Population Prospects. The 2004 Revision. New York, Department of Economic and Social Affairs, Population Division, 2005.
4 Barker WW, Luis CA, Kashuba A, Luis M, Harwood DG, Loewenstein D, Waters C, Jimison P, Shepherd E, Sevush S, Graff-Radford N, Newland D, Todd M, Miller B, Gold M, Heilman K, Doty L, Goodman I, Robinson B, Pearl G, Dickson D, Duara R: Relative frequencies of Alzheimer disease, Lewy body, vascular and frontotemporal dementia, and hippocampal sclerosis in the state of Florida brain bank. Alzheimer Dis Assoc Disord 2002;16:203–212.
5 Fratiglioni L, Launer LJ, Andersen K, Breteler MM, Copeland JR, Dartigues JF, Lobo A, Martinez-Lage J, Soininen H, Hofman A: Incidence of dementia and major subtypes in Europe: a collaborative study of population-based cohorts. Neurologic diseases in the elderly research group. Neurology 2000;54: S10–S15.
6 Alzheimer A: Über eine eigenartige Erkrankung der Hirnrinde. Allg Z Psychiatr Psychisch Gerichtl Med 1907;64:146–148.
7 Ziegler-Graham K, Brookmeyer R, Johnson E, Arrighi HM: Worldwide variation in the doubling time of Alzheimer's disease incidence rates. Alzheimers Dement 2008;4:316–323.
8 Corrada MM, Brookmeyer R, Paganini-Hill A, Berlau D, Kawas CH: Dementia incidence continues to increase with age in the oldest old: the 90+ Study. Ann Neurol 2010;67:114–121.
9 Alzheimer's Disease International: World Alzheimer Report, 2010.
10 Wimo A, Winblad B, Jonsson L: The worldwide societal costs of dementia: estimates for 2009. Alzheimers Dement 2010;6:98–103.
11 Prince M: Dementia in developing countries. A consensus statement from the 10/66 Dementia Research Group. Int J Geriatr Psychiatry 2000;15: 14–20.
12 Prince MJ: The 10/66 Dementia Research Group – 10 years on. Indian J Psychiatry 2009;51(suppl 1): S8–S15.
13 Kikuchi K, Takahashi R, Sugihara Y, Inagi Y: Five-year experience with the long-term care insurance system in japan. J Am Geriatr Soc 2006;54: 1014–1015.
14 Kumamoto K, Arai Y, Zarit SH: Use of home care services effectively reduces feelings of burden among family caregivers of disabled elderly in Japan: preliminary results. Int J Geriatr Psychiatry 2006;21: 163–170.
15 Brookmeyer R, Gray S, Kawas C: Projections of Alzheimer's disease in the United States and the public health impact of delaying disease onset. Am J Public Health 1998;88:1337–1342.

16 Brookmeyer R, Johnson E, Ziegler-Graham K, Arrighi HM: Forecasting the global burden of Alzheimer's disease. Alzheimers Dement 2007;3: 186–191.

17 Brodaty H, Breteler MM, Dekosky ST, Dorenlot P, Fratiglioni L, Hock C, Kenigsberg PA, Scheltens P, De Strooper B: The world of dementia beyond 2020. J Am Geriatr Soc 2011;59:923–927.

18 McKhann G, Drachman D, Folstein M, Katzman R, Price D, Stadlan EM: Clinical diagnosis of Alzheimer's disease: report of the NINCDS-ADRDA Work Group under the auspices of Department of Health and Human Services Task Force on Alzheimer's Disease. Neurology 1984;34:939–944.

19 Sperling RA, Aisen PS, Beckett LA, Bennett DA, Craft S, Fagan AM, Iwatsubo T, Jack CR, Kaye J, Montine TJ, Park DC, Reiman EM, Rowe CC, Siemers E, Stern Y, Yaffe K, Carrillo MC, Thies B, Morrison-Bogorad M, Wagster MV, Phelps CH: Toward defining the preclinical stages of Alzheimer's disease: recommendations from the National Institute on Aging-Alzheimer's Association workgroups on diagnostic guidelines for Alzheimer's disease. Alzheimers Dement 2011;7:280–292.

20 Jack CR Jr, Albert M, Knopman DS, McKhann GM, Sperling RA, Carrillo MC, Thies B, Phelps CH: Introduction to revised criteria for the diagnosis of Alzheimer's disease: National Institute on Aging and the Alzheimer's Association workgroup. Alzheimers Dement 2011;7:257–262.

21 McKhann GM, Knopman DS, Chertkow H, Hyman BT, Jack CR Jr, Kawas CH, Klunk WE, Koroshetz WJ, Manly JJ, Mayeux R, Mohs RC, Morris JC, Rossor MN, Scheltens P, Carillo MC, Thies B, Weintraub S, Phelps CH: The diagnosis of dementia due to Alzheimer's disease: recommendations from the National Institute on Aging and the Alzheimer's Association workgroup. Alzheimers Dement 2011;7: 263–269.

22 Albert MS, Dekosky ST, Dickson D, Dubois B, Feldman HH, Fox NC, Gamst A, Holtzman DM, Jagust WJ, Petersen RC, Snyder PJ, Carrillo MC, Thies B, Phelps CH: The diagnosis of mild cognitive impairment due to Alzheimer's disease: recommendations from the National Institute on Aging-Alzheimer's Association workgroups on diagnostic guidelines for Alzheimer's disease. Alzheimers Dement 2011;7:270–279.

23 Hardy J, Allsop D: Amyloid deposition as the central event in the aetiology of Alzheimer's disease. Trends Pharmacol Sci 1991;12:383–388.

24 Dubois B, Feldman HH, Jacova C, Cummings JL, Dekosky ST, Barberger-Gateau P, Delacourte A, Frisoni G, Fox NC, Galasko D, Gauthier S, Hampel H, Jicha GA, Meguro K, O'Brien J, Pasquier F, Robert P, Rossor M, Salloway S, Sarazin M, de Souza LC, Stern Y, Visser PJ, Scheltens P: Revising the definition of Alzheimer's disease: a new lexicon. Lancet Neurol 2010;9:1118–1127.

25 De Meyer G, Shapiro F, Vanderstichele H, Vanmechelen E, Engelborghs S, De Deyn PP, Coart E, Hansson O, Minthon L, Zetterberg H, Blennow K, Shaw L, Trojanowski JQ: Diagnosis-independent Alzheimer disease biomarker signature in cognitively normal elderly people. Arch Neurol 2010;67: 949–956.

26 Craig-Schapiro R, Kuhn M, Xiong C, Pickering EH, Liu J, Misko TP, Perrin RJ, Bales KR, Soares H, Fagan AM, Holtzman DM: Multiplexed immunoassay panel identifies novel CSF biomarkers for Alzheimer's disease diagnosis and prognosis. PLoS One 2011;6:e18850.

27 Kaduszkiewicz H, Zimmermann T, Beck-Bornholdt HP, van den Bussche H: Cholinesterase inhibitors for patients with Alzheimer's disease: systematic review of randomised clinical trials. BMJ 2005;331: 321–327.

28 Schneider LS, Dagerman KS, Higgins JP, McShane R: Lack of evidence for the efficacy of memantine in mild Alzheimer disease. Arch Neurol 2011.

29 Schneider LS, Insel PS, Weiner MW: Treatment with cholinesterase inhibitors and memantine of patients in the Alzheimer's disease neuroimaging initiative. Arch Neurol 2011;68:58–66.

30 Pimplikar SW, Nixon RA, Robakis NK, Shen J, Tsai LH: Amyloid-independent mechanisms in Alzheimer's disease pathogenesis. J Neurosci 2010;30: 14946–14954.

31 Wyss-Coray T: Inflammation in Alzheimer disease: driving force, bystander or beneficial response? Nat Med 2006;12:1005–1015.

32 Muller WE, Eckert A, Kurz C, Eckert GP, Leuner K: Mitochondrial dysfunction: common final pathway in brain aging and Alzheimer's disease – therapeutic aspects. Mol Neurobiol 2010;41:159–171.

33 Blacker D, Haines JL, Rodes L, Terwedow H, Go RC, Harrell LE, Perry RT, Bassett SS, Chase G, Meyers D, Albert MS, Tanzi R: ApoE-4 and age at onset of Alzheimer's disease: the NIHM genetics initiative. Neurology 1997;48:139–147.

34 Grossman I, Lutz MW, Crenshaw DG, Saunders AM, Burns DK, Roses AD: Alzheimer's disease: diagnostics, prognostics and the road to prevention. EPMA J 2010;1:293–303.

35 Payami H, Montee K, Kaye J: Evidence for familial factors that protect against dementia and outweigh the effect of increasing age. Am J Hum Genet 1994; 54:650–657.
36 Arvanitakis Z, Wilson RS, Schneider JA, Bienias JL, Evans DA, Bennett DA: Diabetes mellitus and progression of rigidity and gait disturbance in older persons. Neurology 2004;63:996–1001.
37 Akomolafe A, Beiser A, Meigs JB, Au R, Green RC, Farrer LA, Wolf PA, Seshadri S: Diabetes mellitus and risk of developing Alzheimer disease: results from the Framingham Study. Arch Neurol 2006;63: 1551–1555.
38 Messier C, Teutenberg K: The role of insulin, insulin growth factor, and insulin-degrading enzyme in brain aging and Alzheimer's disease. Neural Plast 2005;12:311–328.
39 Alafuzoff I, Aho L, Helisalmi S, Mannermaa A, Soininen H: Beta-amyloid deposition in brains of subjects with diabetes. Neuropathol Appl Neurobiol 2009;35:60–68.
40 Beeri MS, Rapp M, Silverman JM, Schmeidler J, Grossman HT, Fallon JT, Purohit DP, Perl DP, Siddiqui A, Lesser G, Rosendorff C, Haroutunian V: Coronary artery disease is associated with Alzheimer disease neuropathology in APOE4 carriers. Neurology 2006;66:1399–1404.
41 Forstl H, Haass C, Hemmer B, Meyer B, Halle M: Boxing-acute complications and late sequelae: from concussion to dementia. Dtsch Arztebl Int 2010;107: 835–839.
42 Roberts GW, Gentleman SM, Lynch A, Murray L, Landon M, Graham DI: Beta-amyloid protein deposition in the brain after severe head injury: implications for the pathogenesis of Alzheimer's disease. J Neurol Neurosurg Psychiatry 1994;57:419–425.
43 Bazarian JJ, Cernak I, Noble-Haeusslein L, Potolicchio S, Temkin N: Long-term neurologic outcomes after traumatic brain injury. J Head Trauma Rehabil 2009;24:439–451.
44 Guskiewicz KM, Marshall SW, Bailes J, McCrea M, Cantu RC, Randolph C, Jordan BD: Association between recurrent concussion and late-life cognitive impairment in retired professional football players. Neurosurgery 2005;57:719–726.
45 Geerlings MI, den Heijer T, Koudstaal PJ, Hofman A, Breteler MM: History of depression, depressive symptoms, and medial temporal lobe atrophy and the risk of Alzheimer disease. Neurology 2008;70: 1258–1264.
46 Correia SC, Santos RX, Cardoso S, Carvalho C, Santos MS, Oliveira CR, Moreira PI: Effects of estrogen in the brain: is it a neuroprotective agent in Alzheimer's disease? Curr Aging Sci 2010;3:113–126.
47 Vardy ER, Rice PJ, Bowie PC, Holmes JD, Grant PJ, Hooper NM: Increased circulating insulin-like growth factor-1 in late-onset Alzheimer's disease. J Alzheimers Dis 2007;12:285–290.
48 Sherwin BB: Estrogen and cognitive functioning in women. Endocr Rev 2003;24:133–151.
49 Espeland MA, Rapp SR, Shumaker SA, Brunner R, Manson JE, Sherwin BB, Hsia J, Margolis KL, Hogan PE, Wallace R, Dailey M, Freeman R, Hays J: Conjugated equine estrogens and global cognitive function in postmenopausal women: Women's Health Initiative Memory Study. JAMA 2004;291: 2959–2968.
50 Sherwin BB: The critical period hypothesis: can it explain discrepancies in the oestrogen-cognition literature? J Neuroendocrinol 2007;19:77–81.
51 Daniel JM, Bohacek J: The critical period hypothesis of estrogen effects on cognition: insights from basic research. Biochim Biophys Acta 2010;1800: 1068–1076.
52 Henderson VW, Brinton RD: Menopause and mitochondria: windows into estrogen effects on Alzheimer's disease risk and therapy. Prog Brain Res 2010;182:77–96.
53 Snowdon DA: Aging and Alzheimer's disease: lessons from the Nun Study. Gerontologist 1997;37: 150–156.
54 Stern Y: Cognitive reserve and Alzheimer disease. Alzheimer Dis Assoc Disord 2006;20:112–117.
55 Iacono D, Markesbery WR, Gross M, Pletnikova O, Rudow G, Zandi P, Troncoso JC: The Nun Study: clinically silent AD, neuronal hypertrophy, and linguistic skills in early life. Neurology 2009;73: 665–673.
56 Jedrziewski MK, Lee VM, Trojanowski JQ: Lowering the risk of Alzheimer's disease: evidence-based practices emerge from new research. Alzheimers Dement 2005;1:152–160.
57 Valenzuela MJ, Sachdev P: Brain reserve and dementia: a systematic review. Psychol Med 2006;36: 441–454.

Maria C. Carrillo, PhD
Alzheimer's Association
225 N. Michigan Ave, Suite 1700
Chicago, IL 60601 (USA)
Tel. +1 312 335 5722, E-Mail Maria.Carrillo@alz.org

Hampel H, Carrillo MC (eds): Alzheimer's Disease – Modernizing Concept, Biological Diagnosis and Therapy.
Adv Biol Psychiatry. Basel, Karger, 2012, vol 28, pp 15–29

The Genetics of Alzheimer's Disease

Matthew C. Schu[a] · Richard Sherva[a] · Lindsay A. Farrer[a] · Robert C. Green[b]

[a]Biomedical Genetics, Boston University School of Medicine, and [b]G2P Research Program, Partners Center for Personalized Genetic Medicine, Division of Genetics, Department of Medicine, Brigham and Women's Hospital and Harvard Medical School, Boston, Mass., USA

Abstract

As the most common form of dementia, Alzheimer's disease (AD) currently affects nearly 34 million people worldwide, with a prevalence of about 5.4 million cases in the United States alone. Given the rapid expansion of the aging population in developed countries, the global prevalence of AD cases is projected to triple in 40 years. The primary risk factor for AD is age, and it is estimated that roughly 1 in 8 Americans over the age of 65 have the disease. Besides age, there are several environmental factors that appear to influence risk of AD, including diet, physical activity, cognitive activity, obesity, hypertension, diabetes and smoking. However, the disease has long been known to disproportionally affect certain family lineages, and heritability estimates for the disease range between 60 and 80%. Until recently the search for genetic influences affecting AD risk remained a largely frustrating endeavor, with the exception of a few causal variants discovered in pedigrees exhibiting rare autosomal dominant forms of the disease. Current efforts to identify genetic risk factors for AD have been complicated by both clinical heterogeneity, where multiple manifestations of dementia are currently diagnosed as AD; and locus heterogeneity, which occurs when different families have different AD risk variants. Recent genome-wide association studies (GWAS) that have enrolled tens of thousands of case-control subjects have revealed a new wave of genetic markers that may better explain the genetic influences of AD and point to new insights in pathophysiology and therapeutic development.

Early Family Studies of Alzheimer's Disease

Affecting an estimated 34 million people worldwide, Alzheimer's disease (AD) is the most common form of dementia in adults [1–2]. While risk for developing the neurodegenerative disorder increases with age, the disease has long been known to disproportionally affect certain lineages and current heritability estimates for AD range between 60 and 80% [3]. The first genetic insights into the heritability of AD came from studying families with an autosomal dominant inheritance pattern (fig. 1) [4]. In addition to

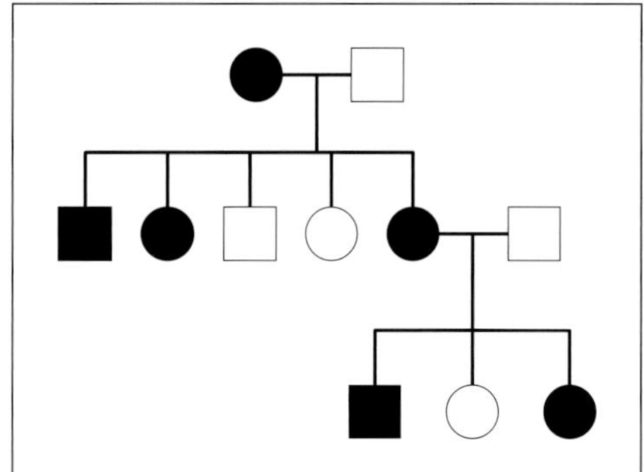

Fig. 1. In an autosomal dominant inheritance pattern the offspring of an affected parent have a 50% risk of developing the disease themselves.

having a clear mendelian inheritance pattern, affected individuals in these pedigrees were noted to manifest disease symptoms at a younger age, often before age 60, and sometimes as early as age 30 [5]. Using a technique known as linkage analysis (which identifies relatively broad chromosomal sections of DNA transmitted from parent to offspring that trend with inheritance patterns of the disease), researchers identified three genes with autosomal dominant mutations segregating with AD cases: amyloid precursor protein *(APP)*, presenilin 1 *(PSEN1)*, and presenilin 2 *(PSEN2)* [6]. Although mutations in these genes collectively account for much less than 1% of all AD cases, our understanding of basic AD pathology was substantially advanced by later studies involving these genes and their protein products [5, 7]. For example, subsequent experiments showed that certain mutations in these genes were associated with increased production of a particular isoform of the protein β-amyloid (Aβ) which has a propensity to aggregate in the intracellular space and eventually form the neuritic plaques first observed by some of the earliest neuropathologists studying AD [5, 8, 9]. The evidence from these and other functional studies surrounding *APP, PSEN1*, and *PSEN2* contributed to the development of the 'amyloid hypothesis' which postulates that the pathogenesis of AD is driven by an imbalance between the production and clearance of Aβ [4]. This theory and the biological roles of each of these genes are discussed below in further detail.

Amyloid Precursor Protein

Perhaps the most important of all the genes that were discovered from early genetic studies of autosomal dominant early-onset Alzheimer's disease (EOAD) is amyloid precursor protein (*APP*). Located on chromosome 21, *APP* encodes a ubiquitous transmembrane protein, which has been postulated to play a role in cell movement and cell adhesion [5, 9]. However, the gene is of particular importance to scientists engaged in AD research as it encodes the protein that is selectively degraded into Aβ. Degradation of APP occurs via two pathways: amyloidogenic and non-amyloidogenic

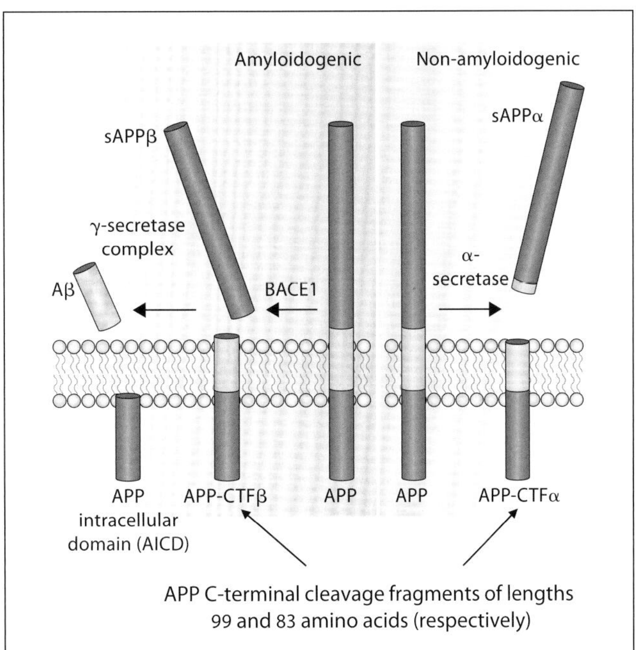

Fig. 2. APP degradation occurring via the amyloidogenic and non-amyloidogenic pathways.

(fig. 2). The amyloidogenic pathway for APP proteolysis is initiated by β-site amyloid precursor protein-cleavage enzyme 1 (BACE-1), resulting in the intermediate product sAPPβ. This segment is then further processed by the complex γ-secretase and the product of this reaction is Aβ [9]. By contrast, in non-amyloidogenic APP proteolysis, an α-secretase first cleaves APP at a different location, 83 amino acids from the C terminus. The location of this cleavage point prevents Aβ production, because it falls within the region of APP that would otherwise be processed into Aβ [9].

Other variables in the processing of APP also affect the formation and toxicity of Aβ monomers. Depending on where the γ-secretase cleaves the sAPPβ subunit, the resulting Aβ molecule will contain either 40 or 42 amino acids. While both of these molecules are present in normally functioning brain tissue, the longer isoform of Aβ is more hydrophobic than the more common $A\beta_{40}$ molecule, and consequently is more prone to aggregation [9, 10]. Autosomal dominant mutations within the *APP* gene have been shown to cause increases in net production of Aβ and elevated ratios of $A\beta_{42}$:$A\beta_{40}$, thus making it more likely for neuritic plaques to form and aggregate [10]. However, it is worth noting that while $A\beta_{42}$ molecules comprise the majority of the monomers found in amyloid plaques of AD patients, it is the small aggregates of oligomers and protofibrils, which may be comprised of either $A\beta_{40}$ or $A\beta_{42}$, that are considered the most neurotoxic forms of Aβ [9]. In particular, these more soluble aggregates of Aβ appear to be highly damaging to synapses, initiating dysfunctional events including (but not limited to) the endocytosis of N-methyl-D-aspartate (NMDA) surface receptors [11].

Table 1. Summary of EAOD attributed mutations

Gene	Mutations, n	Families, n
APP	32	89
PSEN1	185	405
PSEN2	13	22

From the Alzheimer Disease and Frontotemporal Dementia Mutation Database (www.molgen.vib-ua.be/ADMutations).

The Presenilins

Mutations within presenilin 1 and 2 (*PSEN1* and *PSEN2*) were also found to cause autosomal dominant patterns of AD transmission [12, 13]. Located on chromosomes 14 and 1, respectively, the two proteins are highly homologous, and either protein may be found (in conjunction with nicastrin, anterior pharynx defective 1, and presenilin enhancer 2) in the γ-secretase complex which enzymatically cleaves sAPPβ to produce Aβ. Mutations in either gene are also linked to an increased ratio of $A\beta_{42}$:$A\beta_{40}$ monomers [4, 5, 7, 10]. *PSEN1* mutations account for the largest proportion of autosomal dominant AD cases and functional studies associate mutations in *PSEN1* with increased production of $A\beta_{42}$. *PSEN1* has also been implicated in other processes that may be relevant to AD pathogenesis such as notch signaling [5, 7]. Pathological mutations in the *PSEN2* gene represent a much rarer portion of EAOD cases, with many *PSEN2* mutant carriers being related to a family known as the Volga Germans, a well-studied pedigree with an extensive history of AD [7, 13]. As with *PSEN1* mutations, *PSEN2* mutations are also associated with an increased ratio of $A\beta_{42}$:$A\beta_{40}$ in the brain, though mechanisms by which this change occurs are less clear [7].

It should be noted that the mutations in *APP, PSEN1*, and *PSEN2* have been documented in a relatively small number of families with a history of AD, and many mutations have only been observed in one or two pedigrees (table 1).

While the amyloid hypothesis is still at the forefront of many working models of AD, the development of Aβ aggregates is not the only conformational change that has been observed in the brains of AD patients or AD animal models. In addition to Aβ deposits, other changes in the brains of AD patients relative to those of non-demented age-matched controls include increased inflammatory response, oxidative stress, synaptic damage, neuronal death and hyperphosphorylation of tau particles leading to the development of neurofibrillary tangles [9, 11]. These observations coupled with the fact that the majority of AD cases do not follow a mendelian-dominant inheritance pattern are reminders of the need for further genetic explanations for the disease beyond *APP* and the presenilins.

Table 2. Environmental risk factors for AD (modified from Barnes and Yaffe [2])

	Population prevalence, %	Relative risk odds ratio (95% CI)
Worldwide		
Diabetes mellitus	6.4	1.39 (1.17–1.66)
Midlife hypertension	8.9	1.61 (1.16–2.24)
Midlife obesity	3.4	1.60 (1.34–1.92)
Depression	13.2	1.90 (1.55–2.33)
Physical inactivity	17.7	1.82 (1.19–2.78)
Smoking	27.4	1.59 (1.15–2.20)
Low education	40.0	1.59 (1.35–1.86)
USA		
Diabetes mellitus	8.7	1.39 (1.17–1.66)
Midlife hypertension	14.3	1.61 (1.16–2.24)
Midlife obesity	13.1	1.60 (1.34–1.92)
Depression	19.2	1.90 (1.55–2.33)
Physical inactivity	32.5	1.82 (1.19–2.78)
Smoking	20.6	1.59 (1.15–2.20)
Low education	13.3	1.59 (1.35–1.86)

Early Linkage and Candidate Studies of Late-Onset Alzheimer's Disease

Late-onset Alzheimer's disease (LOAD) is by far the more common form of AD and typically is diagnosed in patients older than 65 years of age [5]. In contrast to EOAD, LOAD follows a non-mendelian, more complex inheritance pattern, but there is still a clearly heritable component [14, 15]. There are many challenges to identifying the genetic underpinnings of this more prevalent form of the disease, including inherent clinical heterogeneity of AD, whereby multiple manifestations of dementia are currently diagnosed as the same disease, and locus heterogeneity, which occurs when different families have different AD risk variants that cause a lack of fitness within distinct biological functions but result in the same phenotype in patients. In addition, experiments that are underpowered may fail to detect alleles with small to modest risk effects and are likely to produce false negatives in follow-up analyses [6]. Moreover, epidemiological studies have revealed several well-established environmental and biological risk factors influencing AD onset including patient histories of head trauma, depression, diabetes, level of educational training, physical and cognitive leisure activities, diet, and of course the most influential risk factor, age (table 2) [2, 16, 17].

In spite of these challenges, early linkage studies performed on Caucasian families with a history of LOAD consistently identified a strong linkage peak over the *APOE* region of chromosome 19 as being associated with AD risk [6]. These early linkage studies also identified several other non-*APOE* signals of inherited genetic risk for AD with consistently replicated linkage peaks on chromosomes 6, 9, 10 and 12 that contained many biologically plausible genes which were later proposed for follow-up in candidate gene studies. Despite the strength of some of these signals, only the *APOE* gene on chromosome 19 has been definitively established as a risk factor for LOAD from this group of risk loci identified by linkage and other studies [6, 18, 19].

Moving beyond linkage analysis studies, new insights about AD pathogenesis emerging from cell biology and neurochemistry studies forwarded the advancement of *candidate gene association studies* for AD [8]. These studies tested for disease risk associations among known variants selected from a priori knowledge of the disease by comparing the differences in their allelic frequencies between cases and controls. Because cases and controls did not need to be related, candidate genes studies had the advantage over linkage analyses of not being restricted to family-based data. While these early association studies still lacked power to detect rarer variants influencing AD risk, a few genes did emerge from these efforts that have been replicated across multiple candidate gene studies. For example, there are two loci that have been robustly identified in candidate genes studies, namely neuronal sortilin-related receptor (*SORL1*) and the angiotensin converting enzyme (*ACE*), as well as apolipoprotein E (*APOE*) which was previously identified by linkage studies.

Sortilin-Related Receptor-1

Neuronal sortilin-related receptor (*SORL1*) is a protein trafficking gene that is most likely involved in AD pathogenesis via its interaction with APP. *SORL1* regulates the trafficking of APP from the plasma membrane to the retromer recycling endosome that eventually transports the molecule to the trans-Golgi apparatus (fig. 3) [20]. This processing pathway diverts APP from the amyloidogenic pathway, which reduces the production of Aβ. Genetic variations of *SORL1* are associated with a reduction in function of this trafficking protein and an increased risk for AD. Interestingly, variants at two distinct locations in this gene have been independently shown to influence AD risk. Thus, *SORL1* illustrates that intralocus heterogeneity is another challenge for identifying specific variants of interest in AD [19, 20].

Angiotensin Converting Enzyme

The angiotensin converting enzyme (*ACE*) is responsible for the production of angiotensin II (AT_2) by cleaving angiotensin I (AT_1), and is a well-known target of many antihypertensive drugs. The complete relationship between *ACE* and AD pathogenesis is still unclear, although its association with the disease may suggest a vascular component of AD etiology. It has also been suggested that *ACE* may be capable of cleaving Aβ proteins, offering a link between *ACE* and the amyloid hypothesis. However, this cleavage has only been observed thus far in vitro [21].

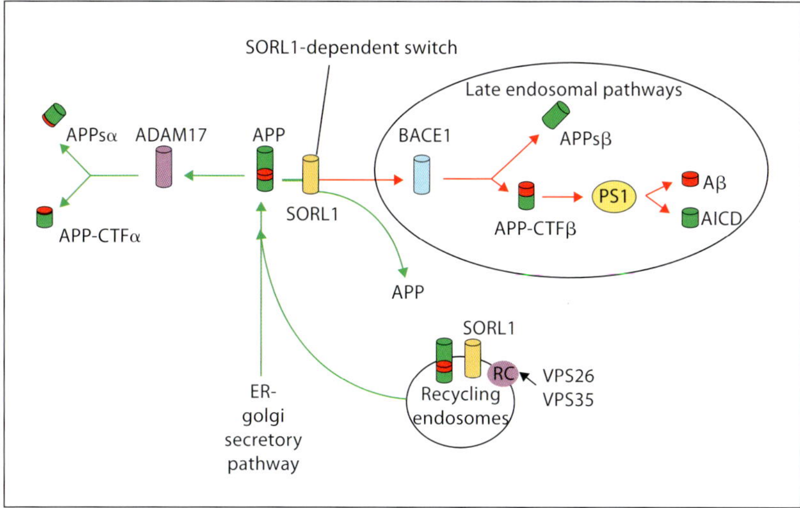

Fig. 3. *SORL1* regulates the trafficking of APP from the plasma membrane to the retromer recycling endosome that eventually transports the molecule to the trans-Golgi apparatus (from Rogaeva et al. [20]).

Apolipoprotein E

Apolipoprotein E (*APOE*) is the most widely studied and robust genetic risk factor for AD. First identified in linkage analysis peaks and then again in subsequent candidate gene studies, it is now known that there are three common isoforms of the gene that differ from each other at amino acid residues 112 and 158. The variants, labeled as ε2, ε3, and ε4 alleles, contain cysteine/cysteine, cysteine/arginine, and arginine/arginine residues and are found in the population at frequencies of 8, 77 and 15% respectively [22]. These substitutions affect both the charge and three-dimensional structure of the ApoE protein, and consequently many of its binding properties with other molecules. ApoE is a major determinant of lipid transport and is expressed throughout the body, though primarily in the liver and brain [22, 23]. Genetic association studies show roughly a 2- to 3-fold increased risk of AD associated with the presence of one ε4 allele and at least a 12-fold increased risk of AD for ε4 homozygotes, though these risk effects appear to vary in different ethnicities [18, 23]. In addition, other studies have shown that ε4 carriers have an earlier age of onset for the disease [24, 25]. Meanwhile, the rarer ε2 allele is thought to be protective for AD [26]. *APOE* genotype is among the more robust of susceptibility variants for common complex diseases and thus *APOE* disclosure has been a valuable model for examining the impact of disclosing such risk variants as described in more detail below [27–32].

The ε4 variant of *APOE* has been associated with nearly every major pathway important to AD pathology, including neurodegeneration and synaptic dysfunction (table 3), although many of these associations have not been consistently observed

Table 3. Effect of APOE genotype in AD pathology (modified from Leoni [22])

Events correlated with APOE genotype	Direction of effect in ε4 carriers
APP expression and production	+
Aβ aggregation and deposition	+
Tangle formation	+
Abnormal tau phosphorylation	+
Neuronal toxicity (by proteinase fragments)	+
Cholesterol efflux and uptake from astrocyte and neurons	–
Aβ processing and elimination	–
LRP1 and LDLR mediated Aβ clearance	–
Cholesterol transport and delivery	–
Anti-inflammatory action	–
Synaptic repair	–
Synaptic plasticity	–
Neurite outgrowth	–

(+) = Event/process is observed to increase in the presence of the ε4 allele.
(–) = Event/process is observed to decrease in the presence of the ε4 allele.

across experiments. However, some of the most robust findings are suggestive of ApoE's involvement in Aβ aggregation and clearance [22, 23]. Supportive evidence comes from studies showing that the ε3 variant has a higher affinity for Aβ than ε4 thereby better facilitating its removal from the extracellular space [33, 34]. While the accumulation of data from follow-up studies presents a highly convincing argument for ApoE's importance in AD pathogenesis, the protein's involvement in a wide variety of brain functions has the unfortunate affect of making its precise mechanistic role in AD more elusive. Furthermore, the ε4 variant of the gene is neither necessary nor sufficient for the development of AD. Thus, while the importance of *APOE* as a risk factor for AD is irrefutable, there are clearly heritable features beyond *APOE* and it does not come close to completely explaining the full genetic heritability of the disease [35].

Genome-Wide Association Studies and Consortia Efforts

While candidate gene studies have used a priori information to investigate hundreds of gene variations that might explain the inherited risk of AD, recent technology has facilitated the development of a more unbiased method to the search for novel genetic loci explaining AD risk. Microarray platforms can now assay over 5 million single nucleotide polymorphisms (SNPs) or genetic variants in each subject, and compare cases and controls in large genome-wide association studies (GWAS). Even given the impressive number of SNPs tested, the expectation of GWAS is not that researchers will directly identify the causal variant(s) behind AD pathology, but rather that the coverage of the chip allows a researcher to detect a signal from SNPs that are in linkage disequilibrium with the causal variants. It is important to note this distinction, as it explains why the results of a GWAS do not provide detailed functional information about a particular variant. Instead, positive results from a GWAS tell us which variants occur at significantly different allelic frequencies between cases and controls, and may point us to the genes surrounding or near these variants.

Simultaneously testing millions of SNPs at a time is a huge technological achievement, but given that most studies enroll *only* a few thousand individuals, the interpretation of GWAS results presents a statistical challenge. To protect against false positive associations that might result from multiple testing, researchers typically impose a highly conservative threshold to declare genome-wide significance for a particular hit (typically a value of less than 5.0×10^{-8}). In recent years, approximately 15 large GWA studies exploring genetic associations for AD risks have been published [19]. While the *APOE* variant was found to be genome-wide significant in all but one of these studies, most of the other genome-wide significant findings were not consistently replicated across these analyses. The lack of fully replicable findings in any particular variant besides *APOE* in GWA studies suggests heterogeneous genetic influences are responsible for a large portion of AD cases [5].

GWAS Consortia

The majority of recently discovered gene associations were derived from large GWAS consortium efforts, where many collaborators studying the genetic influences of AD pooled their data and used meta-analysis of allele frequency patterns to amass greater statistical evidence for these genes' relevance to AD risk. Table 4 lists the top genetic variants (outside of *APOE*), which were described by these consortium efforts [36–41].

Even when a specific gene is identified through consortium efforts, the gene's direct involvement in disease pathology can be difficult to ascertain without functional data available for the protein it encodes. For example, it is suggested that bridging integrator 1 (*BIN1*) may be indirectly linked to AD pathology via its binding partner, dynamin 2, which has previously been associated with AD risk [8, 42]. However, it has also been hypothesized that the products of *BIN1* are involved in receptor-mediated endocytosis and thereby might play a role in the production and/or clearance of Aβ [8]. Similarly, the roles of two membrane-spanning domain proteins, *MS4A4A* and *MS4A6E*, are not clearly understood. However, it is posited that these genes affect signal transduction in the brain given their membership in the multimeric receptor complex [40]. Sometimes there are surprising findings. For example, ephrin type-A receptor 1 (*EPHA1*) has never been linked to a neurodegenerative disorder, but it has been previously implicated in brain cancers, suggesting a role in neuronal growth and development. In addition, a ligand of *EPHA1* (Ephrin-A2) has been associated with the non-amyloidogenic proteolytic pathway of APP [40].

Table 4. Top results from AD consortia

Gene symbol	Non-APOE genes identified	Large AD consortia studies	
CLU	clusterin	• Harold et al. [36] • Seshadri et al. [38] • Hollingworth et al. [39]	• Lambert et al. [37] • Naj et al. [40] • Jun et al. [41]
PICALM	phosphatidylinositol-binding clathrin assembly protein	• Harold et al. [36] • Hollingworth et al. [39] • Jun et al. [41]	• Seshadri et al. [38] • Naj et al. [40]
CR1	complement component receptor 1	• Lambert et al. [37] • Hollingworth et al. [39]	• Naj et al. [40] • Jun et al. [41]
BIN1	bridging integrator 1	• Seshadri et al. [38] • Hollingworth et al. [39]	• Naj et al. [40]
MS4A4A/ MS4A4E6	membrane-spanning 4-domains subfamily A members 4A and E6	• Hollingworth et al. [39]	• Naj et al. [40]
EPHA1	ephrin type-A receptor 1	• Naj et al. [40]	• Seshadri et al. [38]
CD33	myeloid cell surface antigen CD33	• Naj et al. [40]	
CD2AP	CD2-associated protein	• Naj et al. [40]	
ABCA7	ATP-binding cassette transporter-A7	• Hollingworth et al. [39]	

Pathways Consistently Implicated in AD Risk

The majority of the AD-associated genes that have been identified thus far through consortia efforts build upon previous hypotheses regarding AD pathogenesis and add to the body of evidence further implicating pathways such as inflammation, oxidative stress, protein trafficking and Aβ processing with the progression of AD. For example, clusterin (*CLU*) encodes an apolipoprotein which is expressed in the brain, and variants of the gene have been predicted to have pleotropic effects. Rat model experiments show *CLU* co-localizing in neuritic amyloid plaques, and other functional studies reveal evidence that *CLU* may be protective against oxidative stress, damage to cell membranes resulting from inflammation response, apoptosis, and the aggregation of hydrophobic unfolded proteins, like $A\beta_{42}$, all of which have been linked to AD progression [43].

Several of the consortia-identified genes, particularly *CR1* and *CD33*, support evidence of the immune system's involvement in AD. Mouse model studies have shown evidence that complement component receptor 1 (*CR1*) may play a role in clearance of Aβ via the complement system [44]. Meanwhile, the myeloid cell surface antigen *CD33* has been reported to be involved in immune response-induced apoptosis [45].

Endocytosis is a critical process in synaptic transmission and response to neuronal damage, and lack of fitness in cellular endosomal protein trafficking has been linked to several neurodegenerative diseases [40]. Two genes from the collective work of the AD consortia provide strong supporting evidence for the importance of synaptic endocytosis relative to AD progression. The first gene, CD2-associated protein (*CD2AP*), is a scaffold adaptor protein that assists in the regulation of receptor-mediated endocytosis [39]. The second, phosphatidylinositol-binding clathrin assembly protein (*PICALM*), is involved in clathrin-mediated endocytosis, a process whereby several molecules including lipids, growth factors and neurotransmitters are shipped to various parts of the cell for further processing, secretion or degradation. It should also be noted that the associations of *BIN1, CD33* and *SORL1* with AD also support the importance of protein trafficking pathways in AD progression [5, 20].

Finally, the association with ATP-binding cassette transporter A7 (*ABCA7*) appears to strongly support the amyloid hypothesis. *ABCA7* is a transmembrane transporter protein that is highly expressed in the brain. Among other roles, it is involved with APP trafficking and inhibits Aβ production. It is also involved in the clearance of lipids from the cell and has been hypothesized to interact (either directly or indirectly) with *CLU* and *APOE* [39].

Conclusions and Future Directions of Genetic Research

Even with all of the exciting results from recent GWAS collaborations, there is a need for new methods to detect novel risk loci. Investigators have had some early success in applying gene-gene interaction models to tease out epistatic effects between known risk loci and AD progression [41]. Ongoing research regarding the gene-environment interactions that lead to AD progression also promises to elucidate more of the missing heritability thus far unexplained by markers implicated from previous genetic AD association studies [46]. Meanwhile, statistical tools that test for pathway enrichment in lists of the most prominently associated SNPs from GWA results are being developed and tested in other neurological disease datasets and may one day help to decipher broader pathway associations with AD risk [47, 48]. Such pathway insights, along with the discovery of recently identified risk genes, could be influential in the developments in AD treatment by suggesting new targets for drug therapies [49]. Beyond SNP data, researchers are also employing gene expression based technologies to advance our understanding of AD pathogenesis, as evidenced by data published from recent microarray experiments in human neuronal tissue [50] and siRNA knockdown experiments in animal models [51]. In addition, advances in next generation sequence technology (NGS) continue to make whole-exome, whole-genome, and targeted deep sequence screens increasingly more affordable, which will expedite the discovery of the functional variants implicated by known risk markers identified in GWAS and other association studies [8]. Finally, new research investigating the epigenetics of AD promises to elucidate the role which structural changes in the neuronal DNA (i.e. via histone modification or methylation) play in promoting or inhibiting the manifestation of AD given an individual's genotype [52, 53].

Future progress in unraveling the genetic heterogeneity of the AD will likely allow neurologists and other clinicians to offer more effective therapies tailored to the specific genetic profiles of their patients. In fact, some studies have already raised the possibility of different therapeutic effects or side effects based on *APOE* genotype [54–56]. Popular interest in genetic tests for a wide variety of disease risks, including risk of AD, has stimulated the growth of direct-to-consumer genetic testing [57] and a tremendous interest in the impact of receiving genetic risk information. The Risk Evaluation and Education for Alzheimer's Disease Study (REVEAL) is a series of four separate multi-center randomized clinical trials that have collectively enrolled over 1,100 individuals in order to explore emerging themes in the disclosure of genetic information and health outcomes. This work has used the relationship between *APOE* and risk of AD to explore the quantitative development of risk estimates from epidemiological studies and in different ethnic groups [15, 58–60], the emotional impact of disclosing risk information about AD [28, 61, 62], the reasons individuals seek genetic risk information [63, 64], issues in self-perception of genetic risk for AD and how these change with genetic testing [30, 65–67], the degree to which participants recall and value their AD risk assessment results and discuss them with others [68–71], the degree to which genetic testing affects insurance purchasing [27, 72], and the degree to which genetic testing alters health behaviors [29, 73]. In summary, while GWAS consortia have recently interrogated multiple large datasets capable of detecting genes that could have a more subtle effect on AD risk, a genetic explanation for a sizable portion of the heritability of this disease remains to be discovered. However, even at this decidedly incomplete stage, the story of the genetics of AD is already affecting the way in which we understand and respond to this disorder, and pointing the way to new insights and new therapies in the future.

References

1 Alzheimer's Association: Alzheimer's disease facts and figures. Alzheimers Dement 2011;7:208–244.

2 Barnes DE, Yaffe K: The projected effect of risk factor reduction on Alzheimer's disease prevalence. Lancet Neurol 2011;10:819–828.

3 Bienvenu OJ, Davydow DS, Kendler KS: Psychiatric 'diseases' versus behavioral disorders and degree of genetic influence. Psychol Med 2011;41:33–40.

4 Bettens K, Sleegers K, Van Broeckhoven C: Current status on Alzheimer disease molecular genetics: from past, to present, to future. Hum Mol Genet 2010;19(R1):R4–R11.

5 Holtzman DM, Morris JC, Goate AM: Alzheimer's disease: the challenge of the second century. Sci Transl Med 2011;3:77sr71.

6 Ertekin-Taner N: Genetics of Alzheimer disease in the pre- and post-GWAS era. Alzheimers Res Ther 2010;2:3.
7 Bekris LM, Yu CE, Bird TD, Tsuang DW: Genetics of Alzheimer disease. J Geriatr Psychiatry Neurol 2010;23:213–227.
8 Bertram L, Lill CM, Tanzi RE: The genetics of Alzheimer disease: back to the future. Neuron 2010;68:270–281.
9 LaFerla FM, Green KN, Oddo S: Intracellular amyloid-β in Alzheimer's disease. Nat Rev Neurosci 2007;8:499–509.
10 Irvine GB, El-Agnaf OM, Shankar GM, Walsh DM: Protein aggregation in the brain: the molecular basis for Alzheimer's and Parkinson's diseases. Mol Med 2008;14:451–464.
11 Querfurth HW, LaFerla FM: Mechanisms of disease: Alzheimer's disease. N Engl J Med 2010;362: 329–344.
12 Sherrington R, Rogaev EI, Liang Y, et al: Cloning of a gene bearing missense mutations in early-onset familial Alzheimer's disease. Nature 1995;375: 754–760.
13 Levy-Lahad E, Wasco W, Poorkaj P, et al: Candidate gene for the chromosome 1 familial Alzheimer's disease locus. Science 1995;269:973–977.
14 Lautenschlager NT, Cupples LA, Rao VS, et al: Risk of dementia among relatives of Alzheimer's disease patients in the MIRAGE study: what is in store for the oldest old? Neurology 1996;46:641–650.
15 Green RC, Cupples LA, Go R, et al: Risk of dementia among White and African-American relatives of patients with Alzheimer's disease. JAMA 2002;287: 329–336.
16 Plassman BL, Havlik RJ, Steffens DC, et al: Documented head injury in early adulthood and risk of Alzheimer's disease and other dementias. Neurology 2000;55:1158–1166.
17 Akomolafe A, Beiser A, Meigs JB, et al: Diabetes mellitus and risk of developing Alzheimer disease: results from the Framingham Study. Arch Neurol 2006;63:1551–1555.
18 Farrer LA, Cupples LA, Haines JL, et al: Effects of age, sex, and ethnicity on the association between apolipoprotein E genotype and Alzheimer disease. A meta-analysis. APOE and Alzheimer Disease Meta-Analysis Consortium. JAMA 1997; 278:1349–1356.
19 Sherva R, Farrer LA: Power and pitfalls of the genome-wide association study approach to identify genes for Alzheimer's disease. Curr Psychiatry Rep 2011;13:138–146.
20 Rogaeva E, Meng Y, Lee JH, et al: The neuronal sortilin-related receptor SORL1 is genetically associated with Alzheimer disease. Nat Genet 2007;39: 168–177.
21 Miners JS, Baig S, Palmer J, Palmer LE, Kehoe PG, Love S: Aβ-degrading enzymes in Alzheimer's disease. Brain Pathol 2008;18:240–252.
22 Leoni V: The effect of apolipoprotein E (ApoE) genotype on biomarkers of amyloidogenesis, tau pathology and neurodegeneration in Alzheimer's disease. Clin Chem Lab Med 2011;49:375–383.
23 Kim J, Basak JM, Holtzman DM: The role of apolipoprotein E in Alzheimer's disease. Neuron 2009;63: 287–303.
24 Corder EH, Saunders AM, Strittmatter WJ, et al: Gene dose of apolipoprotein E type 4 allele and the risk of Alzheimer's disease in late-onset families. Science 1993;261:921–923.
25 Strittmatter WJ, Saunders AM, Schmechel D, et al: Apolipoprotein E: high-avidity binding to β-amyloid and increased frequency of type 4 allele in late-onset familial Alzheimer's disease. Proc Natl Acad Sci USA 1993;90:1977–1981.
26 Corder EH, Saunders AM, Risch NJ, et al: Protective effect of apolipoprotein E type 2 allele for late onset Alzheimer disease. Nat Genet 1994;7:180–184.
27 Zick CD, Mathews CJ, Roberts J, Cook-Deegan R, Pokorski RJ, Green RC: Genetic testing for Alzheimer's disease and its impact on insurance purchasing behavior. Health Aff (Millwood) 2005; 24:483–490.
28 Green RC, Roberts JS, Cupples LA, et al: Disclosure of APOE genotype for risk of Alzheimer's disease. N Engl J Med 2009;361:245–254.
29 Vernarelli JA, Roberts JS, Hiraki S, Chen CA, Cupples LA, Green RC: Effect of Alzheimer's disease genetic risk disclosure on dietary supplement use. Am J Clin Nutr 2010;91:1402–1407.
30 Linnenbringer E, Roberts J, Hiraki S, Cupples L, Green RC: 'I know what you told me, but this is what I think:' perceived risk of Alzheimer disease among individuals who accurately recall their genetics-based risk estimate. Genet Med 2010;12: 219–227.
31 Roberts J, Christensen K, Green R: Using Alzheimer's disease as a model for genetic risk disclosure: implications for personal genomics. Clin Genet 2011 (E-pub ahead of print).
32 Green R, Lautenbach D: Communicating genetic risk for common disorders in the genomic era. Annu Rev Genomics Hum Genet 2012, in press.
33 LaDu MJ, Pederson TM, Frail DE, Reardon CA, Getz GS, Falduto MT: Purification of apolipoprotein E attenuates isoformt-specific binding to β-amyloid. J Biol Chem 1995;270:9039–9042.
34 Canevari L, Clark JB: Alzheimer's disease and cholesterol: the fat connection. Neurochem Res 2007;32: 739–750.

35 Verghese PB, Castellano JM, Holtzman DM: Apolipoprotein E in Alzheimer's disease and other neurological disorders. Lancet Neurol 2011;10: 241–252.
36 Harold D, Abraham R, Hollingworth P, et al: Genome-wide association study identifies variants at CLU and PICALM associated with Alzheimer's disease. Nat Genet 2009;41:1088–1093.
37 Lambert J, Heath S, Even G, et al: Genome-wide association study identifies variants at CLU and CR1 associated with Alzheimer's disease. Nat Genet 2009;41:1094–1099.
38 Seshadri S, Fitzpatrick AL, Ikram MA, et al: Genome-wide analysis of genetic loci associated with Alzheimer disease. JAMA 2010;303:1832–1840.
39 Hollingworth P, Harold D, Sims R, et al: Common variants at ABCA7, MS4A6A/MS4A4E, EPHA1, CD33 and CD2AP are associated with Alzheimer's disease. Nat Genet 2011;43:429–435.
40 Naj AC, Jun G, Beecham GW, et al: Common variants at MS4A4/MS4A6E, CD2AP, CD33 and EPHA1 are associated with late-onset Alzheimer's disease. Nat Genet 2011;43:436–441.
41 Jun G, Naj A, Beecham G, et al: Meta-analysis confirms CR1, CLU, and PICALM as Alzheimer's disease risk loci and reveals interactions with APOE genotypes. Arch Neurol 2010;67:1473–1484.
42 Aidaralieva NJ, Kamino K, Kimura R, et al: Dynamin-2 gene is a novel susceptibility gene for late-onset Alzheimer disease in non-APOE-ε4 carriers. J Hum Genet 2008;53:296–302.
43 Calero M, Rostagno A, Matsubara E, Zlokovic B, Frangione B, Ghiso J: Apolipoprotein J (clusterin) and Alzheimer's disease. Microsc Res Tech 2000;50: 305–315.
44 Wyss-Coray T, Yan F, Lin AH, et al: Prominent neurodegeneration and increased plaque formation in complement-inhibited Alzheimer's mice. Proc Natl Acad Sci USA 2002;99:10837–10842.
45 Vitale C, Romagnani C, Puccetti A, et al: Surface expression and function of p75/AIRMt-1 or CD33 in acute myeloid leukemias: engagement of CD33 induces apoptosis of leukemic cells. Proc Natl Acad Sci USA 2001;98:5764–5769.
46 Chouliaras L, Sierksma AS, Kenis G, et al: Gene-environment interaction research and transgenic mouse models of Alzheimer's disease. Int J Alzheimers Dis 2010;pii:859101.
47 Holmans P, Green EK, Pahwa JS, et al: Gene ontology analysis of GWA study datasets provides insights into the biology of bipolar disorder. Am J Hum Genet 2009;85:13–24.
48 Yaspan BL, Bush WS, Torstenson ES, et al: Genetic analysis of biological pathway data through genomic randomization. Hum Genet 2011;129:563–571.
49 Sleegers K, Lambert JC, Bertram L, Cruts M, Amouyel P, Van Broeckhoven C: The pursuit of susceptibility genes for Alzheimer's disease: progress and prospects. Trends Genet 2010;26:84–93.
50 Simpson JE, Ince PG, Shaw PJ, et al: Microarray analysis of the astrocyte transcriptome in the aging brain: relationship to Alzheimer's pathology and APOE genotype. Neurobiol Aging 2011;32:1795–1807.
51 Senechal Y, Kelly PH, Cryan JF, Natt F, Dev KK: Amyloid precursor protein knockdown by siRNA impairs spontaneous alternation in adult mice. J Neurochem 2007;102:1928–1940.
52 Twine NA, Janitz K, Wilkins MR, Janitz M: Whole transcriptome sequencing reveals gene expression and splicing differences in brain regions affected by Alzheimer's disease. PLoS One 2011;6:e16266.
53 Balazs R, Vernon J, Hardy J: Epigenetic mechanisms in Alzheimer's disease: progress but much to do. Neurobiol Aging 2011;32:1181–1187.
54 Poirier J, Delisle MC, Quirion R, et al: Apolipoprotein E4 allele as a predictor of cholinergic deficits and treatment outcome in Alzheimer disease. Proc Natl Acad Sci USA 1995;92:12260–12264.
55 Risner ME, Saunders AM, Altman JF, et al: Efficacy of rosiglitazone in a genetically defined population with mild-to-moderate Alzheimer's disease. Pharmacogenomics J 2006;6:246–254.
56 Salloway S, Sperling R, Gilman S, et al: A phase 2 multiple ascending dose trial of bapineuzumab in mild to moderate Alzheimer disease. Neurology 2009;73:2061–2070.
57 Frueh FW, Greely HT, Green RC, Hogarth S, Siegel S: The future of direct-to-consumer clinical genetic tests. Nat Rev Genet 2011;12:511–515.
58 Bachman DL, Green RC, Benke K, Cupples LA, Farrer LA: Comparison of Alzheimer's disease risk factors in White and African-American families: the MIRAGE Study. Neurology 2003;60:1372–1374.
59 Cupples LA, Farrer L, Sadovnick D, Relkin N, Whitehouse P, Green RC: Estimating risk curves for first-degree relatives of patients with Alzheimer's disease: the REVEAL Study. Genet Med 2004;6: 192–196.
60 Christensen KD, Roberts JS, Royal CDM, et al: Incorporating ethnicity into genetic risk assessment for Alzheimer's disease: the REVEAL Study experience. Genet Med 2008;10:207–214.
61 Roberts J, Cupples L, Relkin N, Whitehouse P, Green RC: Genetic risk assessment for adult children of people with Alzheimer's disease: the Risk Evaluation and Education for Alzheimer's Disease (REVEAL) Study. J Geriatr Psychiatr Neurol 2005; 18:250–255.

62 Chung W, Chen C, Cupples L, et al: A new scale measuring psychologic impact of genetic susceptibility testing for Alzheimer disease. Alzheimer Dis Assoc Disord 2009;23:50–56.
63 Roberts JS, LaRusse SA, Katzen H, et al: Reasons for seeking genetic susceptibility testing among first-degree relatives of people with Alzheimer's disease. Alzheimer Dis Assoc Disord 2003;17:86–93.
64 Roberts JS, Barber M, Brown TM, et al: Who seeks genetic susceptibility testing for Alzheimer's disease? Findings from a multisite, randomized clinical trial. Genet Med 2004;6:197–203.
65 LaRusse S, Roberts J, Marteau T, et al: Genetic susceptibility testing versus family history-based risk assessment: impact on perceived risk of Alzheimer disease. Genet Med 2005;7:48–53.
66 Hiraki S, Chen C, Roberts J, Cupples L, Green RC: Perceptions of familial risk in those seeking a genetic risk assessment for Alzheimer's disease. J Genet Couns 2009;18:130–136.
67 Christensen K, Roberts J, Uhlmann W, Green RC: Changes to perceptions of the pros and cons of genetic susceptibility testing after APOE genotyping for Alzheimer disease risk. Genet Med 2011;13:409–414.
68 Eckert SL, Katzen H, Roberts JS, et al: Recall of disclosed apolipoprotein E genotype and lifetime risk estimate for Alzheimer's disease: the REVEAL Study. Genet Med 2006;8:746–751.
69 Ashida S, Koehly LM, Roberts JS, Hiraki S, Green RC: Disclosing the disclosure: factors associated with communicating the results of genetic susceptibility testing for Alzheimer's disease. J Health Commun 2009;14:768–784.
70 Ashida S, Koehly LM, Roberts JS, Chen CA, Hiraki S, Green RC: The role of disease perceptions and results sharing in psychological adaptation after genetic susceptibility testing: the REVEAL Study. Eur J Hum Genet 2010;18:1296–1301.
71 Kopits I, Chen C, Roberts J, Uhlmann W, Green R: Willingness to pay for genetic testing for Alzheimer's disease. Genet Test Mol Biomarkers 2011;15:871–875.
72 Taylor DH Jr, Cook-Deegan RM, Hiraki S, Roberts JS, Blazer DG, Green RC: Genetic testing for Alzheimer's and long-term care insurance. Health Aff (Millwood) 2010;29:102–108.
73 Chao S, Roberts JS, Marteau TM, Silliman RA, Green RC: Health behavior changes after genetic risk assessment for Alzheimer disease: the REVEAL Study. Alzheimer Dis Assoc Disord 2008;22:94–97.

Robert C. Green, MD, MPH
Partners Center for Personalized Genetic Medicine, Division of Genetics
Department of Medicine, Brigham and Women's Hospital and Harvard Medical School
41 Avenue Louis Pasteur, Suite 301, Boston, MA 02115 (USA)
Tel. +1 617 264 5834, E-Mail rcgreen@genetics.med.harvard.edu

Hampel H, Carrillo MC (eds): Alzheimer's Disease – Modernizing Concept, Biological Diagnosis and Therapy.
Adv Biol Psychiatry. Basel, Karger, 2012, vol 28, pp 30–48

Current Conceptual View of Alzheimer's Disease

Karl Herrup

Department of Cell Biology & Neuroscience, Rutgers University, Piscataway, N.J., USA

Abstract

The dementia known as Alzheimer's disease (AD) has been recognized as a clinical entity for nearly 100 years. The neurodegeneration that occurs during the course of the disease is complex, progressive, extensive and defies easy description. Over the years, many different hypotheses have been put forward to explain the underlying biology of AD. While the most widely known of these is the amyloid cascade hypothesis, other models have also received consideration. These alternative views recognize the role of tau, the failure of lysosomal function, abnormalities in autophagy, the loss of cell cycle control, the failure of calcium homeostasis, the establishment of a chronic inflammation and so forth. There are strengths and weaknesses to each of these conceptualizations of the disease, and this review attempts to summarize them. The chapter ends with a proposal that, befitting its complexity, the most productive way to view AD may be as a true composite of all of these mechanisms.

Alzheimer's disease (AD) is arguably the most complex neurodegenerative disease to afflict the human brain. By the final stages of the disease, affected individuals are unable to interact with their surroundings or care for themselves in any meaningful way. Functional deficits appear in virtually every domain of cognition and behavior. At a structural level there are losses of neuronal processes and cell bodies in diverse regions from brainstem to olfactory bulb. These are described in more detail in the following chapter by Schneider and colleagues [pp. 49–70] as are the accumulations of abnormal deposits of many different proteins, which are also scattered widely throughout the brain. Compounding this picture of brain devastation, AD is relentlessly progressive and remarkably common. As detailed in the previous chapter, estimates of prevalence in individuals over age 85 are widely cited to be greater than 50%. And yet, in the vast majority of all cases, no overt signs of impending brain illness are apparent before the age of 60. This is a huge puzzle for anyone seeking to understand the biological basis of AD as it seems nearly unimaginable that a brain that has performed normally for 60+ years could fall apart so completely during a 10- to 20-year span of time. Finally, AD is almost uniquely human. None of the most common experimental animal model systems naturally develops anything resembling the catastrophic degeneration that is seen in AD.

The challenges we face in attempting to conceptualize such a common, age-restricted, nearly uniquely human illness, are immense. Fortunately, we have made substantial progress in recent years in unraveling the mysteries of this modern 'plague', but as of this date it seems fair to say that we have still not achieved consensus on the answer to two deceptively simple questions, 'What is AD and where does it come from?' The goal of this chapter is to review the status of our various attempts to answer these questions.

The Biology of Aging

AD does not strike the young. There are no reported cases of infantile or juvenile-onset AD; even the most virulent cases of familial AD do not appear until the fourth decade of life. Thus, to fully conceptualize AD, we must first understand what is meant by brain aging. Progress in our understanding of the biology of the aging process itself has been remarkable. And this progress has led us to the realization that aging is deeply tied to the most fundamental of all of life processes – nutrition.

In the 1980s, researchers were exploring the possibility that age might be a regulated phenomenon. If so, they reasoned there should be genes whose mutation would extend the typical lifespan of an organism. Using the emerging model organism *Caenorhabditis elegans*, they discovered a single recessive mutation that could extend the life of a wild-type nematode by nearly 50% [1]. They called the mutation *Age1*, and its discovery set off a search for other such genes. From these pioneering studies, a unifying principle emerged. Most 'longevity' genes have functions that relate to the insulin/IGF receptor pathway, which regulates the energy utilization of the cells of the body [2]. The cells of the brain become aware of a rise in serum insulin or insulin-like growth factor (IGF-II) levels through the insulin/IGF-II receptor. Ligand binding initiates a series of signals through the PI3 kinase pathway that resonate through a diversity of critical pathways. One is the stress response pathway. PI3 kinase activates Akt, which then signals through Foxo1 and SIRT1. When translocated to the nucleus, the result is an enhancement in synthesis of stress response genes. A second pathway targeted by insulin-induced PI3 kinase is one of the cell's core signaling proteins, mTOR (mammalian target of rapamycin) [3]. The mTOR protein is another member of the PI3 kinase family and its downstream targets include potent regulators of the protein translation system – again a central function of a cell's biochemistry. mTOR also represses the autophagy pathway in part by phosphorylating the Atg13 protein.

The linkage of this pathway to aging was unexpected at first. The smooth response of the insulin/IGF-II receptor would seem relevant only to a moment-to-moment regulation of energy utilization, yet researchers who study the aging process were repeatedly confronted with the observation that the rate of aging in an organism is highly correlated with the rate of flux through this pathway. The most dramatic demonstration of this relationship is that in organisms from yeast to mammals, caloric restriction (which reduces flux through the insulin/IGF receptor pathway), increases lifespan by as much as twofold. Increasing flux through the pathway, by contrast, accelerates the aging process. To date, this relationship is effective only over a relatively small range; restricting food intake is not the key to immortality. From the vantage point of an age-related disease such as AD, however, the key consideration is that these 'aging' pathways are intertwined with the most fundamental processes of life. And as the risk of developing AD increases with age, we must consider that the origins of AD itself are linked to many of these same processes.

Aging cannot be 'cured'. We must therefore accept that human beings are destined to live with a progressive increase in the risk of developing AD throughout their lives. But while it is linked to aging, AD is not normal aging. It is a pathological process that diverges from the normal biology of brain and body. We must recognize that the task of preventing AD becomes more and more difficult with each year of life; but if we can understand the biology of AD and how it links to the biology of the aging process, we presumably can intervene to either delay its onset or blunt its effects on our nervous system.

Current Theories of Alzheimer's Disease

The time and resources – public and private – that have been put towards uncovering the causes of AD are immense, rivaling those that have been brought to bear on the War on Cancer and the Human Genome Project. As a result, there is no dearth of ideas as to the causes of AD. Some of these theories are more widely accepted than others, but as of this writing, none has been disproven. This begs the question of whether AD might best be viewed as the sum of many mechanisms – some known and probably some that have yet to be discovered. This is an unsatisfying answer to the questions, 'What is AD and where does it come from?' But it is fully consistent with the prevalence of the disease, and with its complex biology and pathological appearance. The following sections briefly summarize several of the better known views of AD.

Alzheimer's Viewed as a Proteinopathy

This is the historical root of our understanding of the disease. In 1906, Alois Alzheimer published the first description of a particularly aggressive form of senile dementia. He took the time in his description to include his results using the newly evolving techniques of heavy metal staining. He reported two types of unusual deposits, which we now know as senile plaques and neurofibrillary tangles. From the correlation of these odd brain lesions with the dementia he concluded that the deposits themselves possibly caused the disease. As neurologists and neuropathologists explored this correlation, they developed an appreciation that the presence of visible tangles and plaques were correlated with disease, but they served as imperfect guides to the sites of neuronal loss. The neurofibrillary tangles of hyperphosphorylated tau were better correlated with the sites of neurodegeneration over a wider range of areas, and with the associated behavioral changes. This period of clinicopathological correlation received a biochemical boost when Glenner and Wong [4] identified the main plaque constituent as macromolecular aggregates of a short peptide that was termed β-amyloid (Aβ). During this same period, other laboratories reported that the tangles that had been identified by Alzheimer consisted of paired helical filaments whose main constituent was a hyperphosphorylated form of the microtubule-associated protein, tau [5, 6]. These discoveries offered clear and compelling disease mechanisms that led from defects in protein processing, to the accumulation of abnormal intra- and extracellular deposits, to the blockage of normal neuronal functioning and ultimately to the atrophy and death of neurons.

Amyloid

After the discovery of the sequence of the Aβ peptide, research soon showed that Aβ was derived from the transmembrane portion of a previously unknown type I transmembrane protein. Because of its relationship to the amyloid-containing plaques, this new protein was named the amyloid precursor protein (APP). When a mutation of the APP gene was subsequently identified as an autosomal dominant form of early-onset AD, the convergence of biochemistry, histopathology and genetics offered nearly overwhelming proof that the key to AD was amyloid. Additional genetic evidence strongly supported this case (see below). The conceptualization of AD as a β-amyloid disease was crystallized in what has become known as the amyloid cascade hypothesis (fig. 1) [7, 8].

According to this hypothesis, AD has two possible initiation points. For familial (genetic) forms of AD, the disease begins with a genetic predisposition to incorrectly process the APP gene, leading to a relentless accumulation of Aβ, in particular its most amyloidogenic form, the 42-amino-acid peptide known as $A\beta_{1-42}$. For sporadic (late-onset) forms of AD, the disease begins with a slow, age-related accumulation of extracellular Aβ that may be influenced by many factors, but has no specific proximal cause. Eventually, in both early- and late-onset forms of AD, this Aβ load reaches a tipping point that starts the amyloid cascade. Single Aβ peptides begin to associate, first in dimers and small molecular weight oligomers. Gradually the oligomers grow in size until microscopically visible plaques form. As the aggregates grow in size a series of neurodegenerative events begins to unfold. Monomeric Aβ peptide has synaptic activity and likely functions as a neuromodulator during normal neuronal activity. The smaller Aβ oligomers – dimers to 12-mers – also have synaptic activity [9, 10];

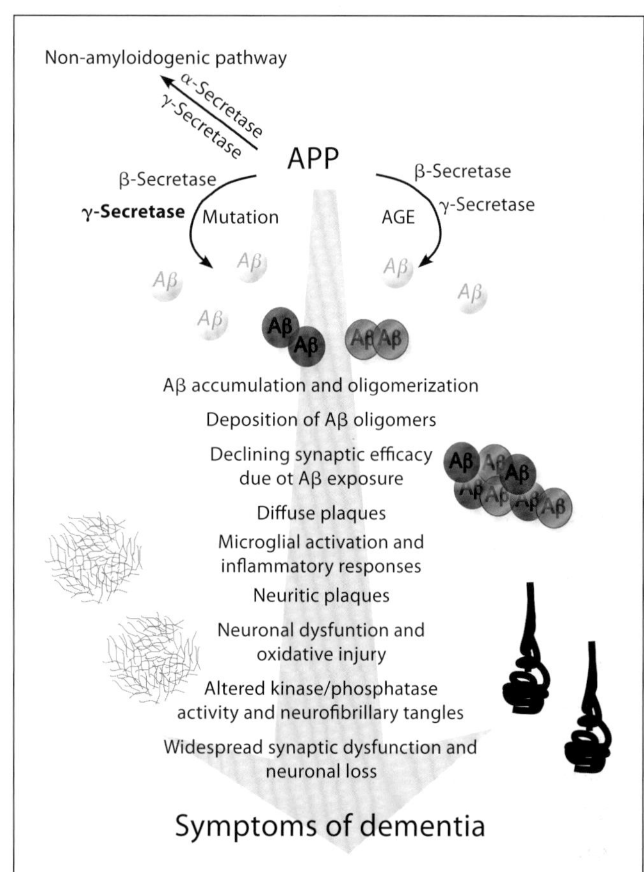

Fig. 1. The amyloid cascade hypothesis [7, 8].

but they have been found to have potent neurotoxic properties as well [11–14]. As the amyloid cascade continues, the aggregating Aβ reaches macromolecular proportions and thus the first plaques appear. Additional pathogenic events add to the destruction: the immune system responds, creating a chronic inflammatory state; neuronal dysfunction moves beyond the synapse leading to the hyperphosphorylation of tau which leads to neuronal cytoskeletal defects and problems with axonal transport. The end result of this multifaceted degenerative cascade is widespread system failure and the progressive symptoms of AD.

Strengths and Weaknesses

The amyloid cascade hypothesis is, by any measure, the most cited conceptualization of the pathogenesis of AD. It has a number of strong points that account for its preeminence. It incorporates both the genetics and the pathology of AD with a single clear program of events. It successfully predicts the reproduction of the memory deficits that are observed in APP-based mouse models of familial AD. It is broad-based and incorporates the known changes in related processes such as tau hyperphosphorylation, oxidative damage and neuroinflammation. Finally, it has offered a lexicon of terms that have helped researchers from around the world to discuss the disease in a consistent fashion. It is not without its critics, however. A conceptual weakness in the hypothesis is that it is largely silent on the mechanisms that lead to the slow accumulation of amyloid that is a required part of the model of sporadic AD. An additional, more powerful criticism is the finding of significant amyloid deposition in individuals with little or no cognitive impairment. According to the amyloid cascade hypothesis, once Aβ begins to aggregate in the brain of an individual and the tipping point is reached, the clinical symptoms of AD should become apparent. But cross-sectional studies, performed either at autopsy or in living patients with plaque imaging, have repeatedly shown that between a quarter and a third of apparently healthy, non-demented individuals carry significant plaque burdens [15, 16]. The mouse models of familial AD carry a similar message. A variety of transgenes – alone or in combination – have been described that produce heavy Aβ plaque deposition in mouse cortex. Despite the fact that the mice carry this amyloid burden for three quarters or more of their lives, none of the models shows significant neurodegeneration or behavioral problems beyond deficits of long-term memory and spatial orientation that do not capture the human condition with any degree of fidelity. The lack of correlation between the anatomical sites of plaque deposition in human brain and neurodegeneration is a related weakness, as is the lack of correlation between plaque burden and cognitive performance. Finally, the results of clinical trials based on this hypothesis have been discouraging. The failure to show efficacy is widely agreed to be due to the fact that by selecting subjects for study who were already afflicted with mild dementia or cognitive impairment, the studies were all begun too late in the disease process to be effective. Current articulations of the model, however, make no prediction that this should be the case.

Tau

AD can also be seen from the vantage point of the neurofibrillary tangle – the second of the abnormal brain deposits identified in Alzheimer's initial report. Conceptualizing AD as primarily a tauopathy is consistent with the amyloid cascade hypothesis in that it views the underlying pathogenic mechanism as a problem of protein misfolding and aggregation. For all of its prominence as a component of the neurofibrillary tangle, tau has no specific hypothesis, equivalent to the amyloid cascade hypothesis, which describes its complete role in the etiology of AD. The general thesis is that the changes of age and disease lead to a faulty regulation of the levels of tau phosphorylation, which alter its ability to interact with its normal protein-binding partners, primarily microtubules. These failed interactions are envisioned as part of the process that drives AD forward. It is a consistent observation that the level of phosphorylation is a key regulatory change that alters the affinity of tau for the microtubule; more heavily phosphorylated forms of tau have weaker affinity for microtubules. As tau binding stiffens the microtubules, reductions in tau/tubulin association will lead to more a more pliable and flexible cytoskeleton. This flexibility is required during certain cellular processes (mitosis and early developmental movements are prime examples), but in the context of a stable adult neuron, the loss of structural integrity and smooth functioning axonal transport can be destructive.

In addition to this loss of function, the enhanced propensity for phosphorylated tau to form aggregates of varying sizes adds a complimentary toxic gain of function as there is evidence that macromolecular assemblages of tau are not good for neurons. As with Aβ, there is evidence that it is the smaller, oligomeric, structures that are more toxic than the larger tangle-forming aggregates [17–19]. Tau-based theories of AD have also advanced in recent years by a series of new discoveries. For example, tau is required for the neurotoxicity observed in Aβ/APP AD models [20, 21]. Further, unlike APP-based mouse models, transgenic animals carrying mutant forms of tau can develop a significant neurodegeneration phenotype [22, 23]. *Drosophila* models make this point particularly well [24]. It is also increasingly apparent that tau has functions beyond those of a simple microtubule-associated protein. For example, the association with the Src-family kinase, Fyn, has broadened the view of the normal function of a tau [25]. Tau hyperphosphorylation increases its affinity for Fyn offering opportunities for impact on functional as well as structural properties of the neurons.

Strengths and Weaknesses

The strengths of the cell biological connections between the tau protein and the health and survival of the neuron are much more direct than with β-amyloid. The anatomical correlations between hyperphosphorylation of tau and the appearance of neurodegeneration in AD are tight and occur early in the disease process [26]. Transgenic mice carrying human tau, both mutant and wild-type develop neuronal cell loss in addition to the hyperphosphorylation phenotype [23], and a similar neurobiology is at work in the fruit fly. Despite these observations, there are weaknesses in using tau as the key to the conceptualization of AD. The most important of these is that the genetics are not in agreement. Not only have mutations in *MAPT* (tau) not been associated with AD, the mutations in *MAPT* that do lead to neurological illness are found in association with a different disorder – FTDP-17 (frontotemporal dementia with parkinsonism linked to chromosome 17) [27]. A second major weakness is that while the tau transgenics will develop hyperphosphorylation and pre-tangle pathology, true neurofibrillary tangles are rare and none of the models develops any degree of plaque pathology. A final weakness facing any theory that is reliant on tau to explain AD is that there is no clear mechanistic explanation for how the disease initiates. Several tau kinases including Cdk5 and GSK3β have been identified and implicated as causative factors in AD, but no generally accepted explanation exists as to what triggers these kinases to malfunction in the temporal and anatomical pattern that leads to AD.

AD can be broken down into two basic types, based on the age at which the patient first demonstrates symptoms. AD is considered late onset when it occurs after the age of 65. This form is usually sporadic in nature, with no specific cause identified. The early-onset form appears before the age of 65 and, by contrast, is usually associated with a mutation in one of three autosomal dominant AD genes: presenilin-1 (PSEN1), APP and presenilin-2 (PSEN2). This means that for all intents and purposes, early-onset AD is a true genetic disease. The involvement of APP and the catalytic γ-secretase subunit with their link to brain amyloid seems clear. Both types of mutations increase the production of $A\beta_{1-42}$.

The picture surrounding late-onset AD is less clear. The *APOE* gene has been identified as a major risk locus for determining age of onset of AD, with the *APOE4* polymorphism repeated shown to enhance the risk of AD by about fourfold. This effect is additive; homozygous *APOE4* individuals are at nearly twice the risk of developing AD as are E4 heterozygotes. No other locus has been identified that approaches this level of involvement. The role of *APOE* in AD is likely to be multifaceted. As a major serum lipid carrier, it likely contributes to vascular functions whose compromise can be an instigating factor in the development of AD. It has also been found that the clearance of excess Aβ from CSF is heavily dependent on *APOE* function [28]. Reduction in this clearance activity would be expected to increase brain Aβ levels and provide an independent pressure towards AD pathology.

Beyond *APOE*, however, the story of AD genetics has taken time to yield to genetic analysis [29]. Modern techniques of human genetics, coupled with increasingly large sample sizes, have enabled researchers to identify several additional loci that confer enhanced risk of AD on their carriers. Some of the more reproducibly identified genes include clusterin, PI-binding clathrin assembly protein, complement receptor 1, bridging integrator 1, ATP-binding cassette transporter, the ephrin A1 receptor, membrane-spanning 4-domains subfamily A, CD2-associated protein, and sialic acid-binding immunoglobulin-like lectin. Estimates are that nearly three quarters of the risk of late-onset AD is genetic in origin and in the aggregate, these genes are estimated to account for about half of that risk [30]. They appear to define three areas of vulnerability: immune function, cholesterol metabolism and synaptic structure/function.

Strengths and Weaknesses

Viewing early-onset forms of AD as a genetic disease is strongly supported by the data. The *APP, PSEN1* and *PSEN2* genes function as autosomal dominant disease genes with nearly full penetrance. As discussed above, the identification of these genes, coupled with the clear ties of their gene products to the biochemistry of the amyloid plaques, strongly supports the major tenets of the amyloid cascade hypothesis. The genes responsible for late-onset forms of AD have also made contributions to our understanding of the disease, the *APOE* locus in particular. But there are caveats that must be recognized in attempting to use them to piece together a comprehensive picture of the causes of AD. Although genetics has been estimated to account for 60–80% of the incidence of AD [31], the enhanced risk associated with any one allele is relatively modest. This situation is to be expected in a complex disease such as AD, but it limits the insights that can obtained. For example, the three areas of vulnerability listed above had all been previously identified based on non-genetic studies. There is hope that the knowledge of which specific proteins are the key players will advance both our understanding of AD and our approaches to its therapy, but this is not yet the case. One aspect of the analyses to date deserves additional comment: certain genes are notable for their absence. Tau, the β- and α-secretases and known 'aging' genes are curiously missing from current lists of probable AD risk factor genes. The data may be advising us to re-examine our biological models and there is surely wisdom in that advice. It is also important to remember that in the study of diseases of aging, the problem of developmental epistasis (an early action of a gene, blocking the ability to document its importance in a later action) may be subverting our ability to discover the impact of key components of the AD phenotype through genetics alone.

Alzheimer's Viewed as a Disease of Neuroinflammation

The reaction of the immune system to the devastation of the CNS during the progression of AD has long been appreciated. Neuropathological and biochemical analyses have consistently shown a strong correlation between disease and the presence of activated microglial cells and reactive astrocytes; biochemical studies have shown enhanced levels of cytokines [32]. These correlations have led many to propose that it is the inflammatory process itself that unleashes the sequence of events that results in the symptoms of AD [33–36]. This view is consistent with the known neurotoxic properties of the products of inflammation including components of the complement cascade [37–41]. There is good data suggesting that the neurons of the locus coeruleus play a role in this process [42, 43]. Strong additional support for the causative role of the immune system in the appearance of AD comes from multiple epidemiological studies indicating that long-term usage of high levels of certain non-steroidal anti-inflammatory drugs (NSAIDs) can reduce the risk of AD by about 50% [44–46]. The web linking inflammation and amyloid is complex and bidirectional. In the mouse, reducing Aβ load can reduce certain indices of inflammation, but not all [47, 48]. Conversely, reducing inflammation can reduce Aβ. In mouse models of familial AD, carrying human *APP* transgenes with or without additional genetic factors such as *PSEN1* or *MAPT*, long-term administration of NSAIDs in the chow reduces plaque burden [49–51]. Equally important, NSAID treatment also blocks the appearance of early indications of neurodegenerative events in these model systems

[51, 52]. Notwithstanding this evidence for a key role of inflammation in AD, however, prospective clinical trials using a variety of different NSAIDs showed no significant benefit in halting or slowing the progression of disease in individuals with mild dementia.

Strengths and Weaknesses
Conceptualizing AD as the product of a chronic inflammatory process in the brain builds on the solid epidemiological evidence for a protective effect of the NSAIDs. It brings together this epidemiological evidence with the histopathology of astrocytes and microglial cells in the AD brain and strong correlative evidence from multiple mouse models of familial AD. The genetics of AD is also beginning to point to a role for inflammation as indicated by the recent GWAS (genome-wide association study) results cited above. The weaknesses in this construct are the failure of the prospective clinical trials such as ADAPT [53]. There is also no clear mechanism for disease initiation – what event triggers the observed chronic inflammatory events. Related to this problem is the uncertainty about whether inflammation is a merely a response to CNS degeneration, caused by other more proximal factors, or whether it is itself a true driving force behind the disease.

Alzheimer's Viewed as a Disease of Mitochondrial Dysfunction and Oxidative Damage

The biological 'rust' we know as oxidative damage affects virtually every component of the cell from DNA to protein to lipids and carbohydrates. A key site for the generation of the reactive oxygen species that lead to these oxidative reactions is the mitochondrion. Both mitochondrial dysfunction and oxidative damage increase with age and thus it is not surprising that both have been suggested as contributing causes of AD. The mitochondrial theory of aging has a long history; its appeal lies in the observation that mitochondrial DNA damage is repaired less avidly, and by different biochemical mechanisms, than is nuclear DNA. Further, the loss of proper mitochondrial function leads directly to impaired energy management in the cell – a loss of mitochondrial membrane potential and the resulting decrease in proper electron transport and efficient ATP production, leading to the generation of oxygen radicals and the resulting oxidative damage. In addition to the contribution of these processes to aging and disease pathogenesis, Aβ appears to act directly on mitochondria, causing their dysfunction [54].

The proposition that mitochondrial dysfunction and oxidative damage are root causes of AD is a compelling hypothesis that links a common theory of aging to AD, an age-related disease. The main weakness in this perspective is that there it does not account for either the anatomical or temporal specificity. The very commonality of the mechanism makes it hard to imagine what causes the specific spectrum of observed cell losses. Further modern genetic analyses have suggested that

Parkinson's disease is more likely linked to aberrant oxidative mechanisms than is AD. Many of the familial Parkinson's mutations identify mitochondrial proteins, and chemicals such as MPTP and rotenone that can cause a parkinsonian-like syndrome in experimental animals are also known mitochondrial toxins. Such linkages to AD are much less apparent.

Alzheimer's Viewed as a Disease of Excitotoxicity and Calcium Dysregulation

The first changes in AD have been argued to occur at the synapse [13, 55], possibly through the disruptive effects of Aβ isoforms. The hypothesis is that the compromise in normal transmitter receptor function leads to global changes in the state of activation of the cell. Overactivation can kill a neuron through the process known as excitotoxic cell death [56, 57]. One common model to explain this phenomenon is that in overactive synapses, the levels of intracellular calcium rise to abnormally high levels. This in turn triggers the activity of a variety of calcium-dependent proteins, calmodulin-dependent kinases and proteases such as calpain. This widespread alteration in synaptic and dendritic homeostasis then leads to a loss of the structural integrity of synapses and dendrites and, if allowed to continue unchecked, results in the initiation of the apoptotic process and the full degeneration of the nerve cell. Excitotoxic cell death can occur at any age in response to overstimulation such as that found after drug administration or during a severe epileptic event.

The suggestion that activity induced calcium dysregulation and the resulting negative consequences for the neurons contributes to the pathogenesis of AD draws strength from independent observations that the regulation of baseline intracellular calcium levels changes with age. Levels have been shown to rise and the regulation of the release of calcium from intracellular stores (mitochondria and endoplasmic reticulum) become more volatile. Intracellular calcium levels rise as the brain ages and there is increased flux through L-type voltage-gated calcium channels; release from intracellular stores triggered by inositol-(1,4,5)-triphosphate and ryanodine receptors also increases. At the same time the NMDA receptor has been shown to allow less calcium to pass through its channel and there is a general reduction in the ability of intracellular stores to properly buffer calcium during normal cellular activity [58]. The intersection of a reduced ability to regulate the levels of intracellular calcium with age plus the chain of degenerative events that ensues from overactive synapses was an early perspective on AD [59] and although thinking has evolved over the years, the connections to disease and aging remain as a centerpiece of many explanations of the origins of AD [60, 61]. The linkage to amyloid has proven to be a significant one [60, 62, 63] and this would appear to tie in yet another thread from the picture of pathogenesis. As with mitochondrial defects and oxidative damage, one unanswered question is how the specific of symptoms of AD – behavioral and anatomical – emerge from a general process such as calcium homeostasis.

Alzheimer's Viewed as a Loss of Cell Cycle Suppression in Adult CNS Neurons

Nearly all neurons in the adult human CNS derive from precursor cells that undergo their final cell division before the end of the first year of life. If for any reason an adult neuron re-enters a cell cycle process, rather than dividing it begins a process of cell death [64]. Cell cycle re-entry can be forced by genetic means [65, 66] or by exposure to certain toxins, including Aβ [67, 68]. Further, in vitro, if the cell cycle is blocked, death can be prevented [69–71]. These data have led to the hypothesis that in AD ectopic initiation of neuronal cell cycling is a major force driving the observed neurodegeneration. This theory proposes that the problem of neurodegeneration in diseases such as AD is fundamentally a loss of cell cycle regulation in the affected nerve cells. The evidence for cell cycle-induced cell death in AD is quite substantial. Indeed the cell cycle/cell death concept was first proposed to explain the presence of mitotic forms of tau protein in the neurons of patients who had died with AD [72, 73]. Beyond mitotic tau epitopes, a variety of proteins normally found only in actively dividing cells can be found in neurons in at-risk regions of the AD brain [74–79]. Unscheduled neuronal cell cycle activity is found at all stages of disease [78, 80], and the phenomenon has been documented as a true cell cycle process that includes DNA replication [77, 81–83]. In addition, the best evidence suggests that it is the cycling neurons that die, albeit over a protracted time period [74, 78, 80].

This perspective raises the question of what features of the AD brain 'pressure' adult neurons to re-enter what will ultimately be a lethal cell cycle. My colleagues and I have hypothesized that one such feature is the altered cellular environment created by chronic brain inflammation. Inflammation accompanies tissue damage in many brain diseases, but as discussed in greater detail above AD is unique in the intimate association found between chronic inflammation and disease. Many different cell types participate in the AD inflammatory response [38, 39, 84–86] – microglial cells, assisted by the responses of astrocytes and brain vascular endothelial cells [86], in addition to the contribution of the locus coeruleus neurons. The result is a chronic shift in the inflammation status of the brain that stresses neurons. My laboratory has proposed that it is this altered brain biochemistry that induces a neuronal cell cycle. Findings in the mouse models of AD are consistent with this idea. In the R1.40 AD mouse, both the anatomical and temporal appearance of neuronal cell cycle events follow the course of the neuropathology found in human AD [14, 87]. Using the cell cycle events as outcome measures, we have shown that 3 months of NSAIDs treatment can completely block neuronal cell cycling while an immune challenge (i.p. injection of LPS) can advance the timing of its appearance by many months.

Alzheimer's Viewed as a Problem of Lysosomal Function and Autophagy

When damage accumulates in a cell, there are two contributing problems. The first is the damage itself and the factors that bring it about. The second is the failure to repair the damage once it has occurred. The problem can be appreciated by considering that in AD, as in other proteinopathies, protein aggregates/inclusions come with disease, but the subunits from which these aggregates are formed are often common byproducts of otherwise normal processes. The cells normally deal with these byproducts by removing them from their site of origin (as discussed above for ApoE) or degrading them in situ. Viewing AD from this perspective, it seems logical to explore the hypothesis that it is the absence of repair as much as the existence of damage per se that is an important source of disease pathogenesis.

One of the main mechanisms by which a cell deals with the accumulation of damage is a process known as autophagy. Although variants known as chaperone-mediated autophagy or microautophagy play their roles, autophagy typically refers to macroautophagy, the process by which large structures – protein aggregates or spent organelles – are packaged and delivered to the lysosome for degradation and reprocessing. The process involves creating a double-membrane vesicle around the structure to be eliminated and the subsequent fusion of this vesicle with resident lysosomes. Autophagy is strongly upregulated by fasting or starvation because it is normally suppressed by the action of mTOR (through its targeting of the autophagy-related protein, Atg13). It can also be triggered by protein aggregates. Failure of autophagy leads to neuronal death and because of this the link to AD has been recognized for many years [88]. Indeed, defects in the appearance of cellular lysosomes is one of the earliest pathological events in the progression that leads to AD [89], particularly in dystrophic neurites.

The connections between autophagy and AD now include a direct mechanistic link with the demonstration that the same mutations in PSEN1 that alter the processing of APP also disrupt the acidification of the lysosome. The effect is indirect; the trafficking of the vesicular proton pump subunit, VOa1, is diverted due to poor glycosylation. As a result, the v-ATPase is under-glycosylated and susceptible to degradation [90]. The net result, however, is a compromise in the autophagic process.

Alzheimer's Viewed as 'All of the Above'

Considered in the aggregate, the preceding discussion of the potential causes of AD hints at a larger perspective on AD that may prove to be the best way to conceptualize the disease. Consider autophagy. It is linked to the regulation of aging through mTOR, linked to the genetics of AD through PS1 function, linked to the proteinopathies of AD through its role in the clearance of large protein aggregates from inside of the cell, and linked to the process of neuronal cell death. Each of these linkages touches a separate view of AD, but autophagy emphasizes the interactions among them at the level of mechanism. The same is true for amyloid and Aβ deposition. They are linked to inflammation, to cell cycle events, to synaptic function and to tau phosphorylation;

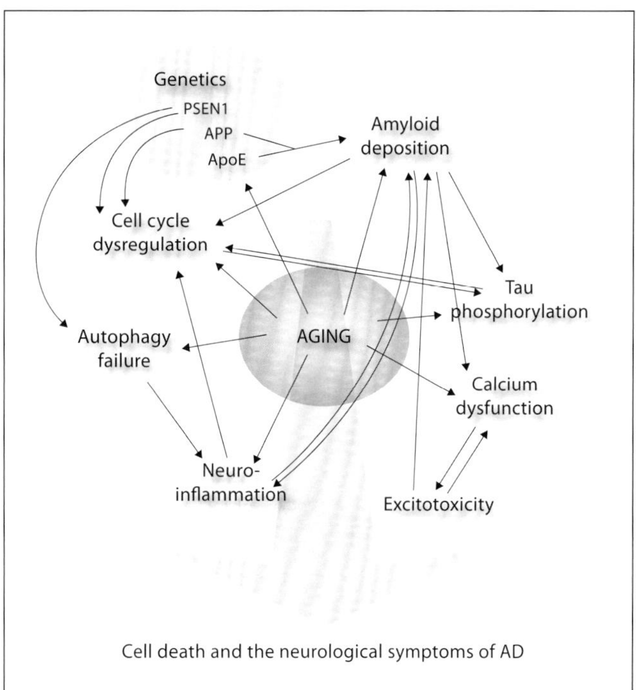

Fig. 2. Interactions of the major causes of AD.

they are impacted by the genetics of early-onset forms of AD and by hyperactivity at the synapse. A similar story applies to tau hyperphosphorylation.

The implication for AD is that at the level of proteins, the problem is a 'network' problem, not a problem with a single pathway. At a cell biological level, AD is not just a neuronal disease; it is a disease of all of the cell types of the brain. This is in no way a new concept. It has been proposed as a unifying idea in diseases such as ALS [91] and others. Major reviews of the AD literature have also begun to adopt this perspective [92, 93], including my own [36]. This interacting web of actions is captured in a small way in figure 2. Much is purposefully left out of this figure, including the specific pathway(s) to the symptoms of AD. The diagram is meant to emphasize the dense nature of the interactions at work in the disease process. Aging is at the hub of the wheel of interactions and impacts them all, but no one process is emphasized more than any other. Unfortunately, while this conceptualization captures the diversity of causes of AD, it is not easily translated into a roadmap for devising therapies or treatments.

My own suggestion, presented previously [36] and summarized in figure 3, is that AD represents a broad-based age-related disease. The essence of this argument is that AD strikes only at an age-weakened brain, but is distinct from normal aging. The emergence of disease represents the end of a three-step process. First, there must be an injury of some sort. This can be a vascular injury, a genetic 'injury', a chemical injury,

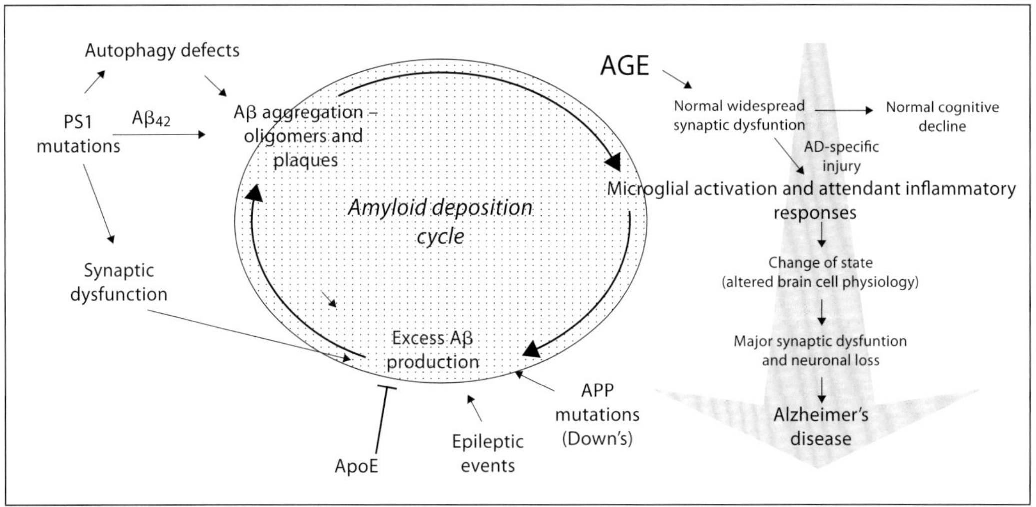

Fig. 3. An age-based model of AD [36].

an infection or any number of other problems. The nature of the injury, however, is less important than the fact that it sets up the second proposed step in the progression of AD, a chronic neuroinflammation based in the aging immune system that changes the cytokine/chemokine chemistry of the brain in an AD-specific way. This long-term inflammatory event then initiates the third step of the process, a cellular change of state. This last step changes the very biology of constituent cells (neurons replicate their DNA, microglial cells enter an alternative activation state, astrocytes adopt an activated phenotype). Some of the individual cellular responses may be reversible, but the state itself is a permanent change in brain biology. The theory maintains that it is from this state that the majority of the degenerative changes leading to cell loss and the symptoms of AD emerge. The production of Aβ and its state of aggregation are prominent players in the process; they form an independent 'amyloid deposition cycle' that interacts closely with the three-step process, but remain mechanistically distinct from the required steps of AD pathogenesis.

Final Perspectives

The strongest message that emerges from a broad, age-based, view of the origins of AD is that the field would benefit from a concerted effort on the parts of industry, government and academia to drop the emphasis on single-pathway origins of disease. The broader view of this disease should stimulate us to adopt a broader range of clinical trials as viable approaches to AD. Anti-inflammatory agents, drugs that counter the hyperphosphorylation or aggregation of tau, treatments to modulate autophagy, methods of blocking the initiation of the neuronal cell cycle – all of these should join with anti-amyloid strategies in searching for ways to improve the lot of those who suffer from or are at risk for developing AD. Aspects of this broad, multifaceted approach are already underway, but we need to do much more. The complexity of the disease in biological terms should prepare us for the likelihood that no one therapy will be successful in all cases; it may not even be completely successful in any case. We should not be deterred by this as we should recognize that altering the 'set' of a network is not a simple task. Nonetheless, the prevalence of AD, its progressive nature, the long-time course between diagnosis and death, and the impact on our society in terms of lost productivity as well as dollars should stiffen our resolve.

References

1 Friedman DB, Johnson TE: A mutation in the age-1 gene in *Caenorhabditis elegans* lengthens life and reduces hermaphrodite fertility. Genetics 1988;118: 75–86.

2 Cohen E, Dillin A: The insulin paradox: aging, proteotoxicity and neurodegeneration. Nat Rev Neurosci 2008;9:759–767.

3 Zoncu R, Efeyan A, Sabatini DM: mTOR: from growth signal integration to cancer, diabetes and ageing. Nat Rev Mol Cell Biol 2011;12:21–35.

4 Glenner GG, Wong CW: Alzheimer's disease: initial report of the purification and characterization of a novel cerebrovascular amyloid protein. Biochem Biophys Res Commun 1984;120:885–890.

5 Grundke-Iqbal I, Iqbal K, Quinlan M, Tung YC, Zaidi MS, Wisniewski HM: Microtubule-associated protein tau. A component of Alzheimer paired helical filaments. J Biol Chem 1986;261:6084–6089.

6 Delacourte A, Defossez A: Alzheimer's disease: tau proteins, the promoting factors of microtubule assembly, are major components of paired helical filaments. J Neurol Sci 1986;76:173–186.

7 Hardy J, Selkoe DJ: The amyloid hypothesis of Alzheimer's disease: progress and problems on the road to therapeutics. Science 2002;297:353–356.

8 Citron M: Strategies for disease modification in Alzheimer's disease. Nat Rev Neurosci 2004;5: 677–685.

9 Wang Q, Walsh DM, Rowan MJ, Selkoe DJ, Anwyl R: Block of long-term potentiation by naturally secreted and synthetic amyloid β-peptide in hippocampal slices is mediated via activation of the kinases c-Jun N-terminal kinase, cyclin-dependent kinase 5, and p38 mitogen-activated protein kinase as well as metabotropic glutamate receptor type 5. J Neurosci 2004;24:3370–3378.

10 Hsieh H, Boehm J, Sato C, Iwatsubo T, Tomita T, Sisodia S, et al: AMPAR removal underlies Aβ-induced synaptic depression and dendritic spine loss. Neuron 2006;52:831–843.

11 Pham E, Crews L, Ubhi K, Hansen L, Adame A, Cartier A, et al: Progressive accumulation of amyloid-β oligomers in Alzheimer's disease and in amyloid precursor protein transgenic mice is accompanied by selective alterations in synaptic scaffold proteins. FEBS J 2010;277:3051–3067.

12 Selkoe DJ: Soluble oligomers of the amyloid β-protein impair synaptic plasticity and behavior. Behav Brain Res 2008;192:106–113.

13 Shankar GM, Li S, Mehta TH, Garcia-Munoz A, Shepardson NE, Smith I, et al: Amyloid-β protein dimers isolated directly from Alzheimer's brains impair synaptic plasticity and memory. Nat Med 2008;14:837–842.

14 Varvel NH, Bhaskar K, Patil AR, Pimplikar SW, Herrup K, Lamb BT: Aβ oligomers induce neuronal cell cycle events in Alzheimer's disease. J Neurosci 2008;28:10786–10793.

15 Villemagne VL, Pike KE, Chetelat G, Ellis KA, Mulligan RS, Bourgeat P, et al: Longitudinal assessment of Aβ and cognition in aging and Alzheimer disease. Ann Neurol 2011;69:181–192.

16 Klunk W, Mathis C, Price J, DeKosky S, Lopresti B, Tsopelas D, et al: Amyloid imaging with PET in Alzheimer's disease, mild cognitive impairment, and clinically unimpaired subjects; in Silverman D (ed): PET in the Evaluation of Alzheimer's Disease and Related Disorders. New York, Springer Science & Business Media LLC, 2009, pp 119–147.

17 Oddo S, Caccamo A, Shepherd JD, Murphy MP, Golde TE, Kayed R, et al: Triple-transgenic model of Alzheimer's disease with plaques and tangles: intracellular Aβ and synaptic dysfunction. Neuron 2003;39:409–421.

18 Spires TL, Orne JD, SantaCruz K, Pitstick R, Carlson GA, Ashe KH, et al: Region-specific dissociation of neuronal loss and neurofibrillary pathology in a mouse model of tauopathy. Am J Pathol 2006; 168:1598–1607.

19 Patterson KR, Remmers C, Fu Y, Brooker S, Kanaan NM, Vana L, et al: Characterization of prefibrillar tau oligomers in vitro and in Alzheimer disease. J Biol Chem 2011;286:23063–23076.

20 Sydow A, Van der Jeugd A, Zheng F, Ahmed T, Balschun D, Petrova O, et al: Tau-induced defects in synaptic plasticity, learning, and memory are reversible in transgenic mice after switching off the toxic Tau mutant. J Neurosci 2011;31:2511–2525.

21 Roberson ED, Scearce-Levie K, Palop JJ, Yan F, Cheng IH, Wu T, et al: Reducing endogenous tau ameliorates amyloid β-induced deficits in an Alzheimer's disease mouse model. Science 2007;316: 750–754.

22 Andorfer C, Acker CM, Kress Y, Hof PR, Duff K, Davies P: Cell-cycle reentry and cell death in transgenic mice expressing nonmutant human tau isoforms. J Neurosci 2005;25:5446–5454.

23 Lee VM, Kenyon TK, Trojanowski JQ: Transgenic animal models of tauopathies. Biochim Biophys Acta 2005;1739:251–259.

24 Wittmann CW, Wszolek MF, Shulman JM, Salvaterra PM, Lewis J, Hutton M, et al: Tauopathy in *Drosophila*: neurodegeneration without neurofibrillary tangles. Science 2001;293:711–714.

25 Roberson ED, Halabisky B, Yoo JW, Yao J, Chin J, Yan F, et al: Amyloid-β/Fyn-induced synaptic, network, and cognitive impairments depend on tau levels in multiple mouse models of Alzheimer's disease. J Neurosci 2011;31:700–711.

26 Braak H, Del Tredici K: The pathological process underlying Alzheimer's disease in individuals under thirty. Acta Neuropathol 2011;121:171–181.

27 Hutton M, Lendon CL, Rizzu P, Baker M, Froelich S, Houlden H, et al: Association of missense and 5′-splice-site mutations in tau with the inherited dementia FTDP-17. Nature 1998;393:702–705.

28 Shibata M, Yamada S, Kumar SR, Calero M, Bading J, Frangione B, et al: Clearance of Alzheimer's amyloid-ss(1-40) peptide from brain by LDL receptor-related protein-1 at the blood-brain barrier. J Clin Invest 2000;106:1489–1499.

29 Bertram L, Lill CM, Tanzi RE: The genetics of Alzheimer disease: back to the future. Neuron 2010;68:270–281.

30 Morgan K: The three new pathways leading to Alzheimer's disease. Neuropathol Appl Neurobiol 2011;37:353–357.

31 Gatz M, Reynolds CA, Fratiglioni L, Johansson B, Mortimer JA, Berg S, et al: Role of genes and environments for explaining Alzheimer disease. Arch Gen Psychiatry 2006;63:168–174.

32 Itagaki S, McGeer P, Akiyama H, Zhu S, Selkoe D: Relationship of microglia and astrocytes to amyloid deposits of Alzheimer's disease. J Neuroimmunol 1989;24:173–182.

33 McGeer PL, Kawamata T, Walker D, Akiyama H, Tooyama I, McGeer E: Microglia in degenerative neurological disease. Glia 1993;7:84–92.

34 McGeer PL, McGeer E: The inflammatory response system of brain: implications for therapy of Alzheimer and other neurodegenerative diseases. Brain Res Rev 1995;21:195–218.

35 McGeer PL, Itagaki S, Tago H, McGeer EG: Reactive microglia in patients with senile dementia of the Alzheimer type are positive for the histocompatibility glycoprotein HLA-DR. Neurosci Lett 1987;79: 195–200.

36 Herrup K: Reimagining Alzheimer's disease – an age-based hypothesis. J Neurosci 2010;30:16755–16762.

37 Akiyama H, Barger S, Barnum S, Bradt B, Bauer J, Cole GM, et al: Inflammation and Alzheimer's disease. Neurobiol Aging 2000;21:383–421.

38 Griffin WS: Inflammation and neurodegenerative diseases. Am J Clin Nutr 2006;83:470S–474S.

39 Griffin WS, Sheng JG, Royston MC, Gentleman SM, McKenzie JE, Graham DI, et al: Glial-neuronal interactions in Alzheimer's disease: the potential role of a 'cytokine cycle' in disease progression. Brain Pathol 1998;8:65–72.

40 Chao CC, Hu S: Tumor necrosis factor-α potentiates glutamate neurotoxicity in human fetal brain cell cultures. Dev Neurosci 1994;16:172–179.

41 Viviani B, Corsini E, Galli CL, Marinovich M: Glia increase degeneration of hippocampal neurons through release of tumor necrosis factor-α. Toxicol Appl Pharmacol 1998;150:271–276.

42 Heneka MT, Ramanathan M, Jacobs AH, Dumitrescu-Ozimek L, Bilkei-Gorzo A, Debeir T, et al: Locus ceruleus degeneration promotes Alzheimer pathogenesis in amyloid precursor protein-23 transgenic mice. J Neurosci 2006;26:1343–1354.

43 Muresan Z, Muresan V: Seeding neuritic plaques from the distance: a possible role for brainstem neurons in the development of Alzheimer's disease pathology. Neurodegener Dis 2008;5:250–253.

44 McGeer PL, Schulzer M, McGeer EG: Arthritis and anti-inflammatory agents as possible protective factors for Alzheimer's disease: a review of 17 epidemiologic studies. Neurology 1996;47:425–432.

45 Stewart W, Kawas C, Corrada M, Metter E: Risk of Alzheimer's disease and duration in NSAID use. Neurology 1997;48:626–632.

46 Vlad SC, Miller DR, Kowall NW, Felson DT: Protective effects of NSAIDs on the development of Alzheimer disease. Neurology 2008;70:1672–1677.

47 Sigurdsson EM, Scholtzova H, Mehta PD, Frangione B, Wisniewski T: Immunization with a nontoxic/nonfibrillar amyloid-β homologous peptide reduces Alzheimer's disease-associated pathology in transgenic mice. Am J Pathol 2001;159:439–447.

48 Morgan D: Modulation of microglial activation state following passive immunization in amyloid depositing transgenic mice. Neurochem Int 2006;49:190–194.

49 Lim GP, Yang F, Chu T, Chen P, Beech W, Teter B, et al: Ibuprofen suppresses plaque pathology and inflammation in a mouse model for Alzheimer's disease. J Neurosci 2000;20:5709–5714.

50 Weggen S, Eriksen JL, Das P, Sagi SA, Wang R, Pietrzik CU, et al: A subset of NSAIDs lower amyloidogenic $A\beta_{42}$ independently of cyclooxygenase activity. Nature 2001;414:212–216.

51 Wilkinson BL, Cramer PE, Varvel NH, Reed-Geaghan E, Jiang Q, Szabo A, et al: Ibuprofen attenuates oxidative damage through NOX2 inhibition in Alzheimer's disease. Neurobiol Aging 2012;33:197.e21–e32.

52 Varvel NH, Bhaskar K, Kounnas MZ, Wagner SL, Yang Y, Lamb BT, et al: NSAIDs prevent, but do not reverse, neuronal cell cycle reentry in a mouse model of Alzheimer disease. J Clin Invest 2009;119:3692–3702.

53 Martin BK, Szekely C, Brandt J, Piantadosi S, Breitner JC, Craft S, et al: Cognitive function over time in the Alzheimer's Disease Anti-inflammatory Prevention Trial (ADAPT): results of a randomized, controlled trial of naproxen and celecoxib. Arch Neurol 2008;65:896–905.

54 Casley CS, Canevari L, Land JM, Clark JB, Sharpe MA: Beta-amyloid inhibits integrated mitochondrial respiration and key enzyme activities. J Neurochem 2002;80:91–100.

55 Selkoe DJ: Alzheimer's disease is a synaptic failure. Science 2002;298:789–791.

56 Wang Y, Qin ZH: Molecular and cellular mechanisms of excitotoxic neuronal death. Apoptosis 2010;15:1382–1402.

57 Szydlowska K, Tymianski M: Calcium, ischemia and excitotoxicity. Cell Calcium 2010;47:122–129.

58 Yu JT, Chang RC, Tan L: Calcium dysregulation in Alzheimer's disease: from mechanisms to therapeutic opportunities. Prog Neurobiol 2009;89:240–255.

59 Khachaturian ZS: Hypothesis on the regulation of cytosol calcium concentration and the aging brain. Neurobiol Aging 1987;8:345–346.

60 Demuro A, Parker I, Stutzmann GE: Calcium signaling and amyloid toxicity in Alzheimer disease. J Biol Chem 2010;285:12463–12468.

61 Supnet C, Bezprozvanny I: The dysregulation of intracellular calcium in Alzheimer disease. Cell Calcium 2010;47:183–189.

62 Bezprozvanny I, Mattson MP: Neuronal calcium mishandling and the pathogenesis of Alzheimer's disease. Trends Neurosci 2008;31:454–463.

63 Green KN, LaFerla FM: Linking calcium to Aβ and Alzheimer's disease. Neuron 2008;59:190–194.

64 Herrup K, Yang Y: Cell cycle regulation in the postmitotic neuron: oxymoron or new biology? Nat Rev Neurosci 2007;8:368–378.

65 Al-Ubaidi MR, Hollyfield JG, Overbeek PA, Baehr W: Photoreceptor degeneration induced by the expression of simian virus 40 large tumor antigen in the retina of transgenic mice. Proc Natl Acad Sci USA 1992;89:1194–1198.

66 Feddersen RM, Clark HB, Yunis WS, Orr HT: In vivo viability of postmitotic Purkinje neurons requires pRb family member function. Mol Cell Neurosci 1995;6:153–167.

67 Kruman II, Wersto RP, Cardozo-Pelaez F, Smilenov L, Chan SL, Chrest FJ, et al: Cell cycle activation linked to neuronal cell death initiated by DNA damage. Neuron 2004;41:549–561.

68 Wu Q, Combs C, Cannady SB, Geldmacher DS, Herrup K: Beta-amyloid activated microglia induce cell cycling and cell death in cultured cortical neurons. Neurobiol Aging 2000;21:797–806.

69 Park DS, Farinelli SE, Greene LA: Inhibitors of cyclin-dependent kinases promote survival of post-mitotic neuronally differentiated PC12 cells and sympathetic neurons. J Biol Chem 1996;271:8161–8169.

70 Park DS, Levine B, Ferrari G, Greene LA: Cyclin-dependent kinase inhibitors and dominant negative cyclin-dependent kinase 4 and 6 promote survival of NGF-deprived sympathetic neurons. J Neurosci 1997;17:8975–8983.

71 Park DS, Morris EJ, Greene LA, Geller HM: G1/S cell cycle blockers and inhibitors of cyclin-dependent kinases suppress camptothecin-induced neuronal apoptosis. J Neurosci 1997;17:1256–1270.

72 Vincent I, Rosado M, Davies P: Mitotic mechanisms in Alzheimer's disease? J Cell Biol 1996;132:413–425.

73 Vincent I, Zheng JH, Dickson DW, Kress Y, Davies P: Mitotic phosphoepitopes precede paired helical filaments in Alzheimer's disease. Neurobiol Aging 1998;19:287–296.

74 Busser J, Geldmacher DS, Herrup K: Ectopic cell cycle proteins predict the sites of neuronal cell death in Alzheimer's disease brain. J Neurosci 1998;18:2801–2807.

75 Hoozemans JJ, Bruckner MK, Rozemuller AJ, Veerhuis R, Eikelenboom P, Arendt T: Cyclin D_1 and cyclin E are co-localized with cyclooxygenase 2 (COX-2) in pyramidal neurons in Alzheimer disease temporal cortex. J Neuropathol Exp Neurol 2002;61:678–688.

76 Nagy Z, Esiri MM, Smith AD: Expression of cell division markers in the hippocampus in Alzheimer's disease and other neurodegenerative conditions. Acta Neuropathol (Berl) 1997;93:294–300.

77 Yang Y, Geldmacher DS, Herrup K: DNA replication precedes neuronal cell death in Alzheimer's disease. J Neurosci 2001;21:2661–2668.

78 Yang Y, Mufson EJ, Herrup K: Neuronal cell death is preceded by cell cycle events at all stages of Alzheimer's disease. J Neurosci 2003;23:2557–2563.

79 Arendt T, Brückner M, Gertz H-J, Marcova L: Cortical distribution of neurofibrillary tangles in Alzheimer's disease matches the pattern of neurons that retain their capacity of plastic remodelling in the adult brain. Neuroscience 1998;83:991–1002.

80 Arendt T, Bruckner M, Mosch B, Losche A: Selective cell death of hyperploid neurons in Alzheimer's disease. Am J Pathol 2010;177:15–20.

81 McConnell MJ, Kaushal D, Yang AH, Kingsbury MA, Rehen SK, Treuner K, et al: Failed clearance of aneuploid embryonic neural progenitor cells leads to excess aneuploidy in the Atm-deficient but not the Trp53-deficient adult cerebral cortex. J Neurosci 2004;24:8090–8096.

82 Mosch B, Morawski M, Mittag A, Lenz D, Tarnok A, Arendt T: Aneuploidy and DNA replication in the normal human brain and Alzheimer's disease. J Neurosci 2007;27:6859–6867.

83 Hoglinger GU, Breunig JJ, Depboylu C, Rouaux C, Michel PP, Alvarez-Fischer D, et al: The pRb/E2F cell-cycle pathway mediates cell death in Parkinson's disease. Proc Natl Acad Sci USA 2007;104:3585–3590.

84 Mucke L, Yu GQ, McConlogue L, Rockenstein EM, Abraham CR, Masliah E: Astroglial expression of human α_1-antichymotrypsin enhances Alzheimer-like pathology in amyloid protein precursor transgenic mice. Am J Pathol 2000;157:2003–2010.

85 Wegiel J, Wang KC, Imaki H, Rubenstein R, Wronska A, Osuchowski M, et al: The role of microglial cells and astrocytes in fibrillar plaque evolution in transgenic APP(SW) mice. Neurobiol Aging 2001;22:49–61.

86 Giri R, Selvaraj S, Miller CA, Hofman F, Yan SD, Stern D, et al: Effect of endothelial cell polarity on β-amyloid-induced migration of monocytes across normal and AD endothelium. Am J Physiol Cell Physiol 2002;283:C895–C904.

87 Yang Y, Varvel NH, Lamb BT, Herrup K: Ectopic cell cycle events link human Alzheimer's disease and amyloid precursor protein transgenic mouse models. J Neurosci 2006;26:775–784.

88 Nixon RA, Yang DS: Autophagy failure in Alzheimer's disease-locating the primary defect. Neurobiol Dis 2011;43:38–45.

89 Nixon RA, Cataldo AM: Lysosomal system pathways: genes to neurodegeneration in Alzheimer's disease. J Alzheimers Dis 2006;9:277–289.

90 Lee JH, Yu WH, Kumar A, Lee S, Mohan PS, Peterhoff CM, et al: Lysosomal proteolysis and autophagy require presenilin 1 and are disrupted by Alzheimer-related PS1 mutations. Cell 2010;141:1146–1158.

91 Ilieva H, Polymenidou M, Cleveland DW: Non-cell autonomous toxicity in neurodegenerative disorders: ALS and beyond. J Cell Biol 2009;187:761–772.

92 Palop JJ, Mucke L: Amyloid-β-induced neuronal dysfunction in Alzheimer's disease: from synapses toward neural networks. Nat Neurosci 2010;13:812–818.

93 Querfurth HW, LaFerla FM: Alzheimer's disease. N Engl J Med 2010;362:329–344.

Karl Herrup, PhD
Department of Cell Biology & Neuroscience, Rutgers University
Piscataway, NJ 08854 (USA)
Tel. +1 732 445 3306
E-Mail herrup@biology.rutgers.edu

Hampel H, Carrillo MC (eds): Alzheimer's Disease – Modernizing Concept, Biological Diagnosis and Therapy.
Adv Biol Psychiatry. Basel, Karger, 2012, vol 28, pp 49–70

Neuropathological Basis of Alzheimer's Disease and Alzheimer's Disease Diagnosis

Julie A. Schneider[a,b] · Thomas J. Montine[c] · Reisa A. Sperling[d] · David A. Bennett[a]

[a]Rush Alzheimer's Disease Center and Department of Neurological Sciences, and [b]Department of Pathology (Neuropathology), Rush University Medical Center, Chicago, Ill., [c]Department of Pathology, University of Washington School of Medicine, Seattle, Wash., and [d]Center for Alzheimer Research and Treatment, Department of Neurology, Brigham and Women's Hospital, Massachusetts General Hospital, Harvard Medical School, Boston, Mass., USA

Abstract

Great strides in the understanding of the neuropathologic basis of Alzheimer's disease (AD) have been achieved over the past century. An overview of the history, current understanding and recent advances in the neuropathologic basis of AD are highlighted in this chapter. Recent advances include the conceptualization and study of the preclinical phase of AD, mild cognitive impairment, mixed pathologies, and the specific roles of β-amyloid and hyperphosphorylated tau protein, as well as other protein accumulations such as soluble amyloid and TDP-43. Furthermore, cerebrospinal fluid and imaging biomarkers in symptomatic and presymptomatic disease are emerging as exciting new tools for both clinical practice and research. These and other advances have led to the formation of new clinical criteria followed by new pathologic criteria for the diagnosis of AD. These advances are likely to have a profound impact on the direction of future research on the prevention and treatment of AD.

From the time Alzheimer's disease (AD) was first described more than a century ago and through to current times, a definite diagnosis of AD has required neuropathologic examination with visualization of the hallmark pathologic changes of senile plaques and neuronal neurofibrillary tangles (NFT). While pathologists continue to use plaques and tangles as the diagnostic features of AD, there have been significant advances in the understanding of the neuropathologic basis of the disease. In this chapter we will discuss the long and fruitful history of clinical-pathologic studies in AD, the neuropathology and current diagnostic criteria for AD, followed by recent advances and new perspectives that are influencing both clinical and research directions in the field. Finally, we will discuss how these and other observations have paved the way for a paradigm shift in the way AD is conceptualized and the amalgamation of new clinical and newly proposed pathologic criteria for the disease.

Historical Perspective: Role of Clinical-Pathologic Studies of Alzheimer's Disease

Loss of cognitive abilities in old age was recognized by the ancient Egyptians 4,000 years ago [1]. However, the clinical description of what we now consider dementia was not established until the 18th century. The French physician Philippe Pinel, considered the father of modern psychiatry, is generally credited for introducing the term dementia (démence), although the term may have been in use prior to that time. However, it was Pinel's student, Jean Etienne Esquirol, in his book *Des Maladies Mentales* (1838), who made the crucial distinction between amentia and dementia, the latter referring to loss of mental faculties leading to the concept of senile dementia [2]. In the 1860s, clinical-pathologic studies of senile dementia demonstrated a reduction of brain weight and atrophy. This was followed by observations relating to hardening of the arteries to senile dementia. Purkinje, who developed crude methods for tissue fixation and staining, is credited for identifying cerebellar neurons (Purkinje cells) in 1827, and for being among the first to use a microtome and a compound microscope in the 1830s [3]. The late 19th century witnessed revolutionary progress in neuropathologic methods including brain tissue fixation, staining, and improved microscope optics. This permitted the characterization of individual neurons in their entirety by Golgi [4] in 1873. These technological developments set the stage for the identification of the lesions that now characterize AD.

Senile plaques were first described by Blocq and Marinesco in 1892 using a carmine stain on brain tissue from patients with chronic epilepsy [5]. However, these scientists did not make a link to dementia. Taking advantage of an improved silver stain developed by Bielschowsky in 1903, Alois Alzheimer published a short report of Auguste D who died at age 55 following a progressive dementia and concluded that her dementia was due to the accumulation of senile plaques and NFT [6]. That same year, Oskar Fischer separately reported a link between dementia and plaques in 12 cases of senile dementia [7]. In his textbook of psychiatry published in 1910, Emil Kraepelin used the term *Alzheimer's disease* to refer to the early onset dementia and distinguished it from the later onset senile dementia [8].

From 1910 until 1985, AD was considered a relatively rare form of dementia affecting people in midlife. It was thought to differ from the much more common senile dementia which was attributed to cerebrovascular disease and the normal aging processes. The reunification of the presenile and senile forms of dementia associated with neuritic plaques and neurofibrillary tangles after three quarters of a century was stimulated in large part by an elegant series of clinical-pathologic studies conducted in the late 1960s by Tomlinson, Blessed and Roth [9–12]. These studies documented a relation between quantitative measures of these hallmark AD lesions and quantitative measures of cognition among persons with dementia. Further, they documented quantitative differences in measures of AD neuropathologic changes and cerebrovascular disease between persons with and without dementia. These findings, confirmed by other groups [13–15], eventually led to the first set of consensus recommendations for the pathologic diagnosis of AD published in 1985 based on the density of senile plaques [16]. These criteria were intended to be used in conjunction with the consensus recommendations for the clinical diagnosis of AD published in 1984 [17]. These and the two following iterations [18, 19] for the neuropathologic diagnosis for AD were 'clinical-pathologic', that is they required a clinical diagnosis of dementia to make a pathologic diagnosis of AD.

Two parallel sets of clinical pathologic studies were unfolding during the 1970s and 1980s that would also have important implications for AD. One was evidence of a relationship between cholinergic deficits in persons with AD to neuronal loss in the basal forebrain, measures of AD pathologic changes, and cognitive function [20–24]. These observations motivated, in part, the conduct of clinical trials with cholinesterase inhibitors [25, 26]. This led to therapeutic interventions approved by the Food and Drug Administration beginning in 1993.

The second observation was the isolation of one of the main protein constituents of senile plaques in 1984, the amyloid-β (Aβ) peptide, initially by Glenner and Wong [27] from cerebrovascular amyloidosis, and subsequently by Masters et al. [28] from neuritic plaque cores. Deposition of Aβ was also shown to be related to cognition [29]. A fuller understanding of the metabolism of Aβ eventually led to two additional important discoveries. First was the development of amyloid imaging agents that could visualize brain amyloid in vivo with positron emission tomography [30, 31]. Second was the identification of therapeutic targets and the development of therapeutic agents with the potential to remove or slow the deposition of amyloid, several of which have gone into human trials. Unfortunately, none of these agents demonstrated safety and efficacy in large clinical trials [32, 33].

Clinical-pathologic studies continue to evolve. Early clinical-pathologic studies focused on characterizing the differences in neuropathologic changes between persons with and without dementia. Although it was clear from these studies that many persons without dementia could have some AD lesions [11, 34, 35], it was not until the past decade that a sufficiently large number of well-characterized individuals without dementia could be the subject themselves of clinical-pathologic investigations. The initial focus was on persons with mild cognitive impairment (MCI) and cognitive impairment no dementia [36, 37]. More recently the focus has been extended to persons without cognitive impairment, i.e. persons without dementia or MCI. Similarly, it has been recognized for many years that AD pathologic changes commonly coexist with other lesions [38, 39]. This phenomenon, also known as mixed pathology, has returned to the forefront of clinical-pathologic study after the observation that these additional pathologies are not only common but are also deleterious in large numbers of community-dwelling older persons [40–45]. Finally, clinical-pathologic studies are now investigating a spectrum of newer proteins as possibly central to the pathogenesis of AD. These more recent advances and perspectives in clinical-pathologic studies on normal aging, MCI, mixed pathologies and additional AD-related proteins will be highlighted later in this chapter.

Overall, clinical-pathologic studies on persons without dementia are shaping our current understanding of the neurobiologic basis of AD. This section highlighted some of the key areas in which these studies informed our understanding of the nosology and natural history of AD and informed on therapeutic targets, some of which resulted in approved therapies, and informed on the development of in vivo biomarkers. We are still at the early stages of understanding AD and clinical-pathologic studies will continue to make important contributions for years and decades to come.

Overview of the Neuropathology of Alzheimer's Disease

Most studies suggest that the lesions of AD accumulate gradually, probably beginning with the widespread accumulation of amyloid plaques followed by NFT [46–52], though the exact sequence is still debated [53, 54]. AD is defined by its microscopic lesions, yet brains with AD also often have macroscopic changes. In general, these changes are neither sensitive nor specific indicators of the disease and cannot be relied on to establish the diagnosis. For instance, brains with AD pathologic changes often have a decrease in weight paralleled by variable cortical atrophy, the latter of which has been correlated with level of cognition [55]. Cortical atrophy in AD is particularly evident in the medial temporal lobe structures, including the hippocampal and entorhinal cortices and with consequent enlargement of the temporal horns of the lateral ventricles. While frontal and parietal regions are also commonly atrophic, the primary cortices, particularly the occipital (primary visual) and perirolandic (primary motor and sensory) cortices, are largely spared. The gross appearance of subcortical structures including the basal ganglia, thalamus and hypothalamus are typically unremarkable, though pigmented structures, especially the substantia nigra and locus cereleus, are often pale in brains with a pathologic diagnosis of AD. Overall, the pattern of atrophy and pallor largely reflects the regions that are commonly and severely affected by the pathology of AD or coexisting pathologic processes. The specific regional pattern of atrophy is noteworthy because these patterns may be a useful as a structural imaging biomarker during life.

The two histological hallmarks of AD are neuronal NFT and the extracellular Aβ deposits of senile plaques. Though NFT are not specific for AD, their numbers are better correlated with level of cognitive impairment compared to amyloid plaque burden [41, 44, 50, 56]. NFT are comprised of abnormally phosphorylated tau protein aggregated as paired helical filaments that occupy the cell body and extend into the apical dendrites. NFT are not easily visualized on hematoxylin and eosin staining, but can be easily visualized by silver impregnation (agyrophilic) methods, such as modified Bielschowsky, Gallyas, Campbell-Switzer, and Bodian stains. Immunohistochemical staining with antibodies to the abnormally phosphorylated tau protein is also a sensitive method for the demonstration of NFT. NFT have a variety of shapes and sizes partially dependent on the size and shape of the host neuron such that cortical NFT are usually flame-shaped or triangular, whereas those in the subcortical or brainstem nuclei are more often globose. In general, NFT also occur in a predictable laminar distribution. In the limbic periallocortex (e.g. entorhinal cortex) NFT are almost always present in large projection neurons of layers II and IV, whereas in the association cortex NFT predominant in layers III and V [57]. In aged brains and in those brains with AD pathology, NFT are observed in a characteristic regional distribution. In 1991, Braak and Braak [48] observed and reported on a characteristic distribution and progression of NFT in the brains of older persons comprising six stages starting in the transentorhinal and entorhinal layers and progressing to the neocortex. In the first two stages there are few NFT mainly in the entorhinal and hippocampal cortices. In stages III and IV there is an increasing accumulation of NFT in the limbic system with few isolated (stage III) or small numbers (stage IV) of NFT in the cortical association areas. Finally in stages V and VI NFT become abundant in both mesial temporal and neocortical association cortices. Braak stage has been related to cognition in multiple studies [39, 41, 44, 45, 50]. Indeed, by current neuropathologic criteria, persons with dementia due to AD must have a Braak III/IV or greater [19]. However, similar to AD pathology as a whole, there is significant overlap. For instance, many older persons without cognitive impairment or a pathologic diagnosis of AD have brains that exhibit a Braak stage of III/IV. Of note, at least some mesial temporal lobe tangles are a near universal finding in the brains of older persons. These NFT in the mesial temporal lobe of older persons may represent an age-related process separate from AD.

Senile plaques, the other microscopic hallmark of AD, consist of fibrillar amyloid material, composed of Aβ protein. Aβ has a characteristic red-green birefringence on Congo red-stained sections of tissue and is produced by the abnormal proteolytic cleavage of amyloid precursor protein (APP). APP is a membrane-bound protein that is normally cleaved by α-secretase to secrete non-amyloidogenic fragments. In AD, Aβ is abnormally cleaved by β- and γ-secretases resulting in the production of Aβ peptide of 39–43 amino acids in length. The insoluble forms of these proteins are deposited primarily as insoluble $A\beta_{40}$ or $A\beta_{42}$ fibrils, with $A\beta_{42}$ often considered the more fibrillogenic and toxic of the species. Plaques are often described as being 'sticky' with other proteins accumulating within the plaques such as interleukins, apoE, and components of the complement system [46]. Similar to NFT, amyloid plaques are typically difficult to identify on routine hematoxylin and eosin stains but are easily identified on silver stains or with protein-specific antibodies. At least two types of plaques accumulate in the AD brain: diffuse plaques (DPs) which are characterized by a focal blush or discoloration without thickened neurites, and neuritic plaques (NPs) which are additionally characterized by thickened neurites, often with a dense central core of amyloid surrounded by a less compact peripheral halo of amyloid. Thickened neurites may contain a variety of different substances, most notably phosphorylated tau protein, which provides a link between NPs and NFT. Unlike DPs which do not appear to disrupt the neuropil, NPs are often associated with focal tissue changes including a reactive astrocytosis and microglial cell activation. Immunostaining of

NPs with antibodies to specific forms of Aβ shows that the dense center core is typically enriched in $A\beta_{40}$, whereas the periphery of the plaque is composed primarily of $A\beta_{42}$ [46]. NPs are considered the more deleterious of the plaques and thus are the plaque type critical for rendering a neuropathologic diagnosis of AD [18, 19]. NPs are prominent in the amygdala and hippocampal subicular complex, and are present in association cortices in AD, but similar to NFT are less common in primary cortices. Some plaques, especially DPs, may have a perivascular orientation, usually in association with intracortical cerebral amyloid angiopathy (CAA). Though some experts have suggested that DPs are early plaques destined to progress to the formation of NPs, some brain regions (i.e. basal ganglia and cerebellar cortex) commonly exhibit DPs without showing significant NP accumulation in spite of long survival times, suggesting DPs and NPs may not just represent two ends of a progression of plaque formation and evolution.

Similar to the stages of NFT as described by Braak and colleagues [46, 47], the progression of senile plaque pathology has also been described. In the first phase, plaques accumulate in neocortical regions of the brain. This is followed by a second phase of plaque deposition in allocortical areas, including the entorhinal cortex and subiculum/CA1 region of hippocampus. Next, in the third phase, amyloid plaque deposition progresses in a caudal direction, accumulating within the basal ganglia, thalamus, and hypothalamus. In the fourth phase, plaques progress into the brainstem accumulating in the midbrain and the medulla oblongata. In the fifth and final phase, senile plaques develop in the pons and cerebellum. There is currently less data on Thal amyloid stages compared to Braak NFT stages, but it is suggested that a stage 4 or greater in amyloid progression is typically associated with dementia, whereas stage 3 is transitional and similar to Braak III/IV may be seen both in persons with and without dementia.

In addition to amyloid deposition in the form of plaques, there is also deposition of Aβ in the leptomeningeal and cortical small arteries and arterioles in nearly all individuals with AD [58]. When severe, CAA is associated with lobar hemorrhages, perivascular scarring and less commonly infarcts. While CAA occurs in all cortical regions, it is most severe in the occipital cortex and overlying meninges. Similar to parenchymal amyloid and in contrast to NFT, the progression of CAA appears to be top down, that is it starts in the neocortex and then progresses to the mesial temporal lobe, followed by the brainstem [59].

Current Diagnostic Criteria for the Neuropathologic Diagnosis of Alzheimer's Disease

Recommended criteria for the pathologic diagnosis of AD were implemented 26 years ago [16] and since that time there have been two modifications [18, 19] and new recommendations are in progress [60]. The first recommendations, the 'Khachaturian criteria' used an age-dependent specific number of senile plaques in a given microscopic cortical field to render a pathologic diagnosis of AD. Older persons were required to have more senile plaques in the microscopic field than younger persons as well as a premorbid diagnosis of dementia. Plaques type was not specified. The Consortium to Establish a Registry for Alzheimer's disease (CERAD) criteria for a pathologic diagnosis of AD refined and simplified the pathologic diagnosis of AD in 1991 [18]. CERAD proposed more easily obtained semiquantitative measures, specifically delineated neocortical NP, and established a probability statement (possible, probable or definite) on the likelihood of AD. Plaques scores were modified by the age at death, and final probability statements were modified by clinical diagnosis. Similar to Khachaturian criteria, probable or definite AD required a larger number of plaques in older persons and a preexisting diagnosis of dementia.

It is important to emphasize that both Khachaturian and CERAD criteria required a greater number or density of plaques in older compared to younger persons to establish a probable or definite diagnosis of AD. For instance, CERAD criteria required only sparse neocortical NPs in a person under the age of 50 years, whereas frequent plaques were required in a person over the age of 75 years for a definite diagnosis. The stipulation that older persons would need more pathology to diagnose a dementia due to AD pathology may now seem counterintuitive. However, the requirement was born out of the observation that small or even moderate amounts of AD pathology were 'normally' observed in older persons without dementia. It was uncertain whether this pathology was 'normal aging' or related to disease. Thus, it was reasoned, since older persons commonly and perhaps 'normally' have some AD pathology, even greater numbers would be necessary for the development of dementia.

The third set of criteria proposed in 1997 which are currently used by most researchers and clinicians are the NIA-Reagan criteria [19]. NIA-Reagan criteria made a couple of important changes to previous criteria. First, in addition to requiring a certain density of NP using CERAD semiquantitative estimates, NIA-Reagan criteria required a specific severity of NFT degeneration using Braak stage [48]. Though NFT are a noted hallmark of AD, they were largely omitted from previous criteria. The inclusion of Braak in the NIA-Reagan criteria was a major advance, however because NIA-Reagan only considered cases where NFT stage paralleled the increasing severity of NP accumulation, it was unclear how to handle the significant proportion of cases with discordant NP and NFT severities. A second pivotal change was that NIA-Reagan criteria did not adjust CERAD plaque estimates with regards to age. The same amount of pathologic change was required for diagnosis whether a person was under 50 or over 75. Removing age from the criteria reflected the changing attitudes regarding the presence of age-related AD lesions. Nonetheless, AD remained a clinical-pathologic diagnosis and similar to CERAD, NIA-Reagan criteria allowed for a probability statement of the likelihood that a dementia was due to AD (high, intermediate, low). It remained unclear how to approach AD pathology in normal aging or MCI. Indeed, research is now focusing on this early phase of clinical disease and preclinical AD pathology leading to new perspectives and prompting the formulation of new pathologic criteria for the diagnosis of AD. These advances and the newest proposed criteria are discussed below.

Recent Advances and New Perspectives

AD Pathology in Cognitively Normal Persons: Preclinical AD

Significant AD pathologic change is commonly observed in older persons without dementia. This was observed in some of the earliest clinical-pathologic studies [34, 35] and has been a continued source of intrigue and debate. Is AD pathologic change an inevitable age-related phenomenon? What are the mechanisms by which AD pathologic changes relate to the clinical expression of dementia? Why would the same amount of AD pathologic change in one person be related to dementia while in another not be related to any evident clinical symptoms? Many of these questions regarding AD pathologic changes in normal older persons are now being answered using prospective longitudinal clinical-pathologic studies and more recently using in vivo cerebrospinal fluid (CSF) and neuroimaging biomarkers. Because these studies perform regular cognitive testing of participants, biomarker collection or autopsy is proximate to cognitive testing. Community-based longitudinal clinical-pathologic studies confirm that significant AD pathologic changes are a relatively common finding in persons without dementia, with about one third of cognitively normal older persons exhibiting sufficient AD pathologic changes to allow a diagnosis of AD if demented [40–45]. On average, this disease burden is less severe than in cases of dementia but there is overlap [61–63].

This concept of preclinical AD pathology has been further solidified in biomarker studies using CSF $A\beta_{42}$ and more directly in vivo brain PET amyloid scanning [64, 65]. These data strongly suggest a high frequency of preclinical AD pathologic change in normal older persons similar to that seen in clinical-pathologic cohorts [66]. In addition, data suggests that these preclinical changes are not static but progress over time [67–69]. Together these clinical-pathologic and biomarker studies have led to the hypothesis that AD begins years to decades prior to the onset of clinical disease.

There is growing evidence to suggest that 'preclinical' AD is not benign. Older persons with overall normal cognitive function who have 'preclinical' AD pathologic change by brain autopsy have lower scores on cognitive functions tests, particularly episodic memory [61, 63]. Aβ biomarker studies are confirming this relationship between preclinical AD and cognition [70–72], moreover recent data are suggesting that these biomarkers may be useful indicators for cognitive decline [73, 74]. Interestingly, at least one clinical-pathologic study indicated that older persons with AD pathologic changes but without dementia are more likely to have memory complaints [75], suggesting that some older persons may recognize these early brain changes and mild memory impairment.

It is likely that older persons have differing interindividual thresholds for the manifestation of cognitive symptoms, i.e. cognitive or neural reserve [76] from AD pathology. Studies are finding that there are medical, genetic, experiential and other factors that increase or decrease the likelihood of clinically expressing AD pathology as dementia [77–82]. There are probably different mechanisms by which these factors impart risk or protection from the expression of AD pathology. One important mechanism is the presence of additional pathologies in a brain already harboring some degree of AD pathology [77–79]. Other mechanisms may include synaptic density which may increase cognitive reserve and cognitive or other activities which may strengthen existing cognitive networks.

While autopsy-based studies can never prove that AD pathologic changes are evidence of preclinical disease (meaning that clinical impairments would have become apparent had the individual lived longer), coupling these cross-sectional observations from autopsy-based studies with results from longitudinal biomarker and neuroimaging studies strongly supports the existence of preclinical AD. Indeed, most diseases have a preclinical phase that can be only be detected using specific screening tools, including blood tests and imaging. For example, many cancers have an early growth phase when there are few if any clinical symptoms. Symptoms develop when the tumor reaches some critical mass and this timing may vary depending on the individual. Indeed, diseases of many organ systems, e.g. cardiac and renal, may be relatively silent until the system is unable to compensate and symptoms develop. While structurally and functionally more complex, the brain is well known to have a great deal of plasticity and it should not come as a surprise the significant pathology would be present prior to the onset of symptoms and that there would be a great deal of individual differences in the extent to which pathology is clinically manifest.

Finally, it is important to recognize that low levels of AD pathologic change are common in the brains of normal older persons that do not fulfill pathologic criteria for the diagnosis of AD (or have biomarker evidence of the same). Indeed, many brains of older persons show sparse AD lesions that do not reach current thresholds for pathologic diagnosis. Data suggests that AD may accumulate slowly over many years, and thus may have a phase with sparse pathology. Very little is known regarding the initiation and early propagation of AD lesions. It is likely that future biomarkers will be able to recognize even earlier stages of disease which may be important for prevention and treatments. Indeed, biomarkers for amyloid may be on the cusp of being able to identify even earlier pathologic changes [83]. Because most of age-related cognitive decline is likely the result of the slow progression of AD and other diseases [84], it will be important to continue to investigate these earliest pathologic stages of the AD.

Neuropathology of Mild Cognitive Impairment

MCI is a clinical diagnosis and represents an intermediate functional stage between normal aging and dementia. Persons with MCI have cognitive impairment, memory or non-memory, but do not fulfill criteria for dementia. In the past decade there has been expanding data on the pathologic basis of MCI [85–92]. Studies suggest that the underlying diseases that result in MCI are heterogeneous, including AD, infarcts and Lewy body disease. AD is the most common pathologic change in persons with MCI, particularly in those with the amnestic subtype [91], and over half of the persons with MCI have sufficient AD pathologic change for pathologic diagnosis [92], suggesting that there is often significantly advanced brain disease at this early stage prior to dementia. Nonetheless, several studies have now shown that on average persons with MCI have an intermediate level of AD pathologic change compared to persons with dementia and no cognitive impairment [89, 90], supporting that MCI represents a transition between normal aging and dementia [89, 92]. Infarcts are also common, particularly in persons with non-amnestic subtype of MCI, and mixed with AD pathology in persons with amnestic MCI. Indeed, mixed pathology, that is AD with infarcts and/or Lewy body disease, may also be present at this early clinical stage [92]. Overall, the heterogeneity and severity of pathology in persons with MCI has implications for preventions and treatments for targeting early disease.

Mixed Pathologies

Older persons often have more than one brain pathology, or mixed pathologies, and data is accumulating that common age-related brain pathologies such as cerebral infarcts and Lewy bodies strongly increase the likelihood of cognitive impairment in persons with AD lesions [40–45]. These additional diseases are likely to have a large public health impact given that longitudinal community-based clinical-pathologic studies suggest dementia is commonly due to mixed pathologies rather than AD pathology alone [93]. Both infarcts and Lewy bodies more commonly coexist with AD pathology then as an isolated disease in older persons with dementia. Indeed, mixed brain pathologies are exceedingly common in the brains of community-dwelling older persons and are more common than any single disease in older persons with dementia. AD pathologic change mixed with infarcts is the most common mixture followed by AD mixed with Lewy bodies. The addition of other diseases is not benign. Some early studies suggested an interaction between AD pathology and infarcts [77]; however, most subsequent studies have suggested an additive effect, such that each additional disease further adds to the likelihood of dementia and severity of cognitive impairment [77–79, 94, 95]. Mixed pathologies are also common in clinically diagnosed probable AD and may be seen in MCI, particularly amnestic MCI, as previously noted [92]. Thus, it is important that both clinicians and researchers recognize mixed pathologies as a central etiology of dementia in older persons. Though AD with infarcts is the most common mixed pathology, AD with Lewy bodies is also common, particularly in clinic or hospital settings where behavioral difficulties may prompt clinic visits or hospitalization [96]. There are other important implications. First, risk factors for a coexisting pathology such as infarcts may appear to be risk factors for AD in clinical studies. For example, a history of diabetes in subjects from a community-based cohort predicted incident clinical AD [97], but an autopsy study from the same cohort found that diabetes was related to infarct rather than AD pathology [98]. Similar misinterpretations of risk factor associations may be occurring in other samples because of unrecognized mixed pathologies. Finally, it is likely that many persons with mixed pathologies are inadvertently included in clinical trials for AD. This has the potential to make it more difficult to find treatment effects specifically directed at amyloid or other specific AD pathologies, although it is possible that it may contribute to putting people on the slippery slope. It is important to carefully consider different ways in which these pathologies could influence the outcome of clinical trials. Ultimately, however, it will be important to direct preventions and treatments toward each of the common age-related pathologies.

Role of Amyloid and Tangles in the Etiopathogenesis of AD

Most would agree that Aβ plaques and NFT are the hallmark lesions of AD, yet their specific role in the pathogenesis of AD continues to be debated [99]. NFT have a stronger association with cognition as compared to amyloid [50, 100], but are less specific, occurring in other diseases and in the mesial temporal lobe of nearly all older individuals. Moreover, mutations involving the tau protein, the major component of NFT, have not been found to result in AD but rather are pathogenic for non-AD neurodegenerative diseases. These and other data suggest NFT are pivotal in progression of AD but probably later in the pathologic cascade of AD [52]. By contrast, there is compelling data to suggest that Aβ is central to the early pathogenesis of AD [52, 101]. First, mutations in the APP gene which encode the APP result in familial autosomal dominant AD. In addition, Down's syndrome which is caused by trisomy 21, the chromosome that contains the APP gene, is associated with inevitable AD pathologic brain changes, and familial cases of AD secondary to mutations of presenilin or APP increase levels of Aβ protein. Furthermore, the presenilins may act as secretases the central enzymes in the abnormal proteolytic cleavage of APP. Finally, considerable data suggest that *APOE* leads to clinical disease in part through an effect on amyloid fibrillogenesis or clearance [101]. Overall, there is compelling data supporting the role of Aβ and NFT in the progression of AD. Yet, there is also new evidence that other pathologic factors are also likely to play a role. Two possible factors are discussed below.

Soluble Amyloid

Insoluble fibrillar Aβ deposition in the form of senile plaques has only a moderate correlation with cognition and may be present in significant amounts in cognitively normal persons. This has led some to posit that something other than the fibrillary Aβ deposition must be related to at least some of the cognitive deficits in AD. One current hypothesis is that soluble species of Aβ may play role in the early pathogenesis and exert an important direct toxicity in AD. If directly toxic, soluble Aβ species may explain the only moderate correlation between senile plaques and cognition. In mouse models of AD, cognitive deficits have been directly related to low amounts of naturally assembled Aβ oligomers, particularly trimers [102]. Interestingly, the cognitive deficits were rapid and transient. One study found a simultaneous rise in nonamers and dodecamers of soluble amyloid, particularly a 56-kDa protein ($A\beta^*_{56}$) coincident with the onset of memory impairment [103]. Recent data support a role for multiple compartments [104] and types of soluble Aβ species, with dimers, trimers, tetramers and $A\beta^*_{56}$ all showing biological activity at various concentrations and potency [105]. The exact role of these soluble species in the progression of human AD is still unclear. Soluble species derived from the human brain have been shown to have toxicity to synapses and to impair memory [106]. There have also been reports of associations of serum and CSF-soluble amyloid species to dementia [107–109] suggesting they could play a role as a biomarker of disease. In addition, soluble species have been associated with the phosphorylation of tau [110] suggesting a possible link between soluble Aβ and the formation of NFT. The relationship between soluble Aβ species and reversible memory loss make them a potential therapeutic target. In the mouse, antibodies directed against Aβ were able to reverse cognitive deficits and slow the deposition of Aβ, suggesting a possible role for removing these soluble Aβ species [111]. However, similar antibodies were not successful and even harmful in human trials, possibly related to massive brain immune response [112, 113]. These intriguing data on the role of soluble amyloid in AD, the direct effects on cognition, and the possible therapeutic benefits of antibody therapies will require further study.

TDP-43 Pathology

Recently a novel abnormal protein accumulation was found in the brains of persons with AD separate from Aβ and NFT [114]. This protein, TDP-43, is a cleaved and phosphorylated form of a naturally occurring DNA RNA nuclear binding protein. With disease, the TDP-43 protein translocates from the nucleus to the cytoplasm. These inclusions were initially described as the pathogenic inclusion in a subset of persons with frontotemporal lobar degeneration and/or amyotrophic lateral sclerosis [115]. More recently it has been found that about one quarter to one half of brains in persons with AD also have similar TDP-43 inclusions [116, 117]. These inclusions are predominant in the mesial temporal lobe and to a lesser extent to the neocortex [118] and appear to be especially common in persons with more advanced AD or with hippocampal sclerosis [116]. Though there are few studies, TDP-43 inclusions may be associated with more severe cognitive impairment in persons with AD [44, 118, 119], MCI [119] and more severe atrophy on structural imaging [117]. It is currently unclear whether TDP-43 inclusion formation occurs within the context of the pathogenesis of AD, i.e. a third pathology of AD, or represents a separate coexisting disease in persons with AD, i.e. mixed pathology.

New Directions in Clinical Criteria and Influence on Pathologic Criteria

In 2009, the National Institute on Aging and Alzheimer's Association convened three workgroups to develop revised clinical and research guidelines for the diagnosis of AD. This initiative was prompted by the need to update the existing NINDS-ADRDA criteria for AD [17]. Although these diagnostic criteria have held up well for over a quarter of a century, they were focused solely on the dementia stage of AD. Over the past decade, it has become increasingly recognized that AD is a continuum of disease from a preclinical to a prodromal and finally an overt dementia phase of the illness. In addition, the advent of both imaging and fluid biomarkers that provide evidence regarding the presence of the AD pathophysiological process in the brain needed to be considered in the diagnostic context of all stages of AD. The workgroups developed a common framework for biomarkers, described in the next section. Although the current recommendations for biomarkers are primarily intended for research purposes, it is recognized that biomarkers will likely play a critical role in improving the diagnostic certainty regarding the underlying etiology of cognitive impairment and dementia, and in moving the field towards early detection and treatment of AD. Overall, the recognition of preclinical AD in older persons without symptoms has been one of the most significant advances in AD research. It offers a unique opportunity to study and follow the early pathogenesis of AD and the development of clinical symptomatology. It is likely that it is during the early phases when treatment will be the most effective. Further studies and trials using antemortem biomarkers for AD including PET amyloid imaging and CSF biomarkers of $A\beta_{42}$ and tau species (see below) in persons with no or MCI will be needed to further unravel the sequence of events leading to cognitive impairment and decline. Further study is needed to correlate biomarker data with neuropathologic changes, and confirm the proposed sequence of neuronal injury and degeneration in older persons without cognitive impairment.

Biomarkers and Their Relationship with Alzheimer's Disease Pathology

Though a definite diagnosis of AD still requires histopathologic evaluation, clinical criteria for AD are beginning to incorporate biomarkers into diagnostic algorithms. By definition, biomarkers are objectively measured in vivo indicators of a physiologic or pathologic process associated with disease or health status. Biomarkers can be used for multiple purposes in both clinical and research settings, including aiding in clinical diagnosis, screening for preclinical disease, evaluating risk and prognosis, and monitoring of potential therapies [64]. Using biomarkers as an adjunct to the research clinical diagnosis of AD was proposed by Dubois et al. [120] in 2007. It was recommended that at least one of three biomarkers (structural neuroimaging with MRI, molecular neuroimaging with PET, or CSF analysis of Aβ or tau proteins) needs be positive to support that episodic memory of at least 6 months' duration is due to

AD [121]. Biomarkers have also been incorporated into the recently published NIA criteria for clinical and research diagnosis of clinical AD [122], MCI [123] and prodromal AD [124]. The new NIA criteria divide biomarkers into those associated with Aβ (low CSF $A\beta_{42}$ or accumulation of amyloid on PET imaging) and those associated with neuronal degeneration or injury (high CSF total or phospho tau, decreased uptake in the temporal-parietal regions on ^{18}F-fluorodeoxyglucose (FDG)-PET scan, or temporal-parietal atrophy on MRI) [125]. In the new NIA clinical criteria for MCI due to AD and AD, the biomarkers are considered complimentary whereas biomarkers are essential for establishing diagnosis in prodromal AD.

Ideally, clinical-pathologic-imaging studies would support the use of specific biomarkers for AD. Biomarkers associated with Aβ have the strong advantage of measuring a hallmark of AD in vivo. Clinical-pathologic studies of the most widely studied radioactive tracer, Pittsburgh Compound B ([^{11}C]PIB), have shown excellent correlations between tracer uptake and postmortem AD pathology [126, 127]. In a double-blind phase 3 study of a florbetapir, an ^{18}F-labeled tracer, a positive scan showed excellent sensitivity and specificity for the pathologic diagnosis of AD [31]. It is important to note that brain Aβ can also accumulate in the form of DPs and in cerebral vessels as CAA, both of which can occur in the absence of a pathologic diagnosis of AD, however it is not yet clear how often these lesions would result in false negative study using PET amyloid imaging. CSF levels of $A\beta_{42}$ have shown excellent sensitivity and specificity in clinical studies, but there is little data on levels of CSF amyloid and the accumulation of brain AD histopathologic lesions. Low CSF $A\beta_{42}$ has been associated with positive tracer uptake on PET amyloid imaging and the proportion of cases showing low CSF amyloid levels is similar to PET amyloid imaging providing supportive evidence that CSF amyloid levels reflect the histopathology of the disease [128].

Current AD biomarkers for neuronal degeneration or injury may be more sensitive indicators of cognitive changes in clinical AD. CSF tau levels have been related to atrophy and volumetric measures in some studies [129], and on structural imaging, hippocampal atrophy has been correlated with AD pathology [130]. But these findings are nonspecific and may also be seen in the context of ischemia and other pathologies [131]. Another biomarker for AD is FDG-PET. Regional hypometabolism may be visualized prior to structural changes in AD using FDG-PET [132]. The neuropathology underlying the FDG-PET signal is not known, but it is hypothesized that the changes reflect neuronal and synaptic loss and degeneration in these regions [52].

More clinical-pathologic-radiologic studies are needed to ascertain the link between biomarkers and the pathophysiologic process of AD. Combining different biomarkers may be an important tool to provide more complete information on the underlying disease pathology and severity [133] and to track progression. It remains important to recognize that even highly specific and sensitive biomarkers for AD will not exclude the presence of additional disease, such as ischemic injury, Lewy body disease, TDP-43 accumulation, and hippocampal sclerosis. Biomarkers for these and other disease processes will likely be important to determine whether a dementia is due to AD only or mixed pathologies.

Revised Neuropathologic Diagnostic Criteria under Consideration

New consensus guidelines for the pathological features of AD are in process. Although not completed, preliminary guidelines from the committee have been presented and opened for review and comments [60]. The new guidelines highlight two major points of revision that will be proposed for the neuropathologic diagnosis of AD.

The first major proposed revision to the pathologic criteria is use of the term 'AD neuropathologic change' to reflect the characteristic histopathologic changes of AD that may be present at any point along the clinical continuum from preclinical to MCI to dementia. This pivotal change disentangles the clinical disease from AD lesions observed by histopathologic means and provides a framework for the nomenclature of AD neuropathologic changes that are often observed in individuals with milder symptoms and those who were cognitively intact during life yet at autopsy have clear evidence of AD pathology. We hasten to add that the proposed revised criteria do not advocate simply reporting histopathologic changes, although this is critically important, but also to compose a clinical-pathologic correlation that integrates the clinical, neuroimaging, laboratory, and neuropathologic information for each individual.

The second major proposed revision of the neuropathologic criteria for AD is to provide a structure for neuropathologic classification and reporting of common comorbidities. The criteria acknowledge that AD commonly does not occur in a 'pure' form and specifically allows for the simultaneous diagnosis of AD and other pathologic comorbidities such as Lewy body disease, ischemic injury, and hippocampal sclerosis. The situation is complicated because each of these other diseases also may occur in a 'pure' form without coexisting AD neuropathologic change or as a neuropathologic feature of other diseases. Again, the culmination of the neuropathologic evaluation is proposed to be a clinical-pathologic correlation that will judge, to the extent possible, the contribution of each disease process observed at autopsy to a given patient's clinical course, understanding that such correlations may be limited and in some instances impossible.

There are also other planned changes. The proposed neuropathologic criteria acknowledge that NFT and Aβ accumulation may progress at different rates and provides new guidelines and appropriate nomenclature for these circumstances. Given the increasing importance of preclinical disease and biomarkers, the new criteria also propose an emphasis on the early pathologic features of AD and the use of the more sensitive molecularly specific antibodies for detection of pathology. The new criteria specifically recognize the increasing significance of Aβ in early disease and as a biomarker. Accordingly, the new criteria will propose the parallel assessment of amyloid stage by the method of Thal et al. [46], in addition to the evaluation of CERAD NP density estimates [18] and Braak NFT stage [48].

Overall, the revised criteria recognize the rapid advancements that are occurring in clinical-pathologic and molecular research and the assessment of genetic and environmental risk and biomarkers for AD. Neuropathologic criteria remain the standard to which new approaches are compared. It is likely that in the near future some combination of risk assessment and biomarker quantification will become complementary or surrogate markers for the presence and progression of neuropathologic changes or cognitive impairment.

Future Directions

The neuropathologic basis of AD is complex. Findings from clinical-pathologic studies have been instrumental in shifting and reshaping the focus of research toward preclinical AD, MCI and mixed pathology. Basic science research has expanded the spectrum of AD pathology with exciting work on soluble amyloid proteins and a possible role for TDP-43 pathology. For the first time in the history of AD, it will be possible to identify and follow the progression of AD pathology using in vivo biomarkers. Together, these advances have influenced the new clinical criteria for preclinical AD, MCI due to AD, and AD dementia. They have similarly influenced the development of new criteria for the pathologic reporting and diagnosis of AD neuropathologic changes. Overall, these research advances and new perspectives in AD have been pivotal and are likely to change the future landscape of diagnosis, prevention and treatment of AD.

References

1 Boller F, Forbes MM: History of dementia and dementia in history: an overview. J Neurol Sci 1998;158:125–133.

2 Berchtold NC, Cotman CW: Evolution in the conceptualization of dementia and Alzheimer's disease: Greco-Roman period to the 1960s. Neurobiol Aging 1998;3:173–189.

3 Posner EJ: Evangelista Purkyně (1787–1869). Br Med J 1969;3:107–109.

4 Pannese E: The Golgi stain: invention, diffusion and impact on neurosciences. J Hist Neurosci 1999;8:132–140.

5 Buda O, Arsene D, Ceausu M, Dermengiu D, Curca GC: Georges Marinesco and the early research in neuropathology. Neurology 2009;72:88–91.

6 Goedert M, Ghetti B: Alois Alzheimer: his life and times. Brain Pathol 2007;17:57–62.

7 Goedert M: Oskar Fischer and the study of dementia. Brain 2009;132:1102–1111.

8 Amaducci LA, Rocca WA, Schoenberg BS: Origin of the distinction between Alzheimer's disease and senile dementia: how history can clarify nosology. Neurology 1986;36:1497–1499.

9 Roth M, Tomlinson BE, Blessed G: Correlation between scores for dementia and counts of 'senile plaques' in cerebral grey matter of elderly subjects. Nature 1966;209:109–110.

10 Blessed G, Tomlinson BE, Roth M: The association between quantitative measures of dementia and of senile change in the cerebral grey matter of elderly subjects. Br J Psychiatry 1968;114:797–811.

11 Tomlinson BE, Blessed G, Roth M: Observations on the brains of non-demented old people. J Neurol Sci 1968;7:331–356.

12 Tomlinson BE, Blessed G, Roth M: Observations on the brains of demented old people. J Neurol Sci 1970;11:205–242.

13 Ulrich J, Stähelin HB: The variable topography of Alzheimer type changes in senile dementia and normal old age. Gerontology 1984;30:210–214.
14 Terry RD, Katzman R: Senile dementia of the Alzheimer type. Ann Neurol 1983;14:497–506.
15 Wilcock GK, Esiri MM: Plaques, tangles and dementia. A quantitative study. J Neurol Sci 1982;56: 343–356.
16 Khachaturian ZS: Diagnosis of Alzheimer's disease. Arch Neurol 1985;42:1097–1105.
17 McKhann G, Drachman D, Folstein M, Katzman R, Price D, Stadlan EM: Clinical diagnosis of Alzheimer's disease: report of the NINCDS-ADRDA Work Group under the auspices of Department of Health and Human Services Task Force on Alzheimer's Disease. Neurology 1984;34:939–944.
18 Mirra SS, Heyman A, McKeel D, Sumi SM, Crain BJ, Brownlee LM, Vogel FS, Hughes JP, van Belle G, Berg L: The Consortium to Establish a Registry for Alzheimer's Disease (CERAD). Part II. Standardization of the neuropathologic assessment of Alzheimer's disease. Neurology 1991;41:479–486.
19 National Institute on Aging, and Reagan Institute Working Group on Diagnostic Criteria for the Neuropathological Assessment of Alzheimer's Disease: Consensus recommendations for the postmortem diagnosis of Alzheimer's disease. Neurobiol Aging 1997;18:S1–S2.
20 Perry EK, Perry RH, Blessed G, Tomlinson BE: Necropsy evidence of central cholinergic deficits in senile dementia. Lancet 1977;1:189.
21 Perry EK, Tomlinson BE, Blessed G, Bergmann K, Gibson PH, Perry RH: Correlation of cholinergic abnormalities with senile plaques and mental test scores in senile dementia. Br Med J 1978;2:1457–1459.
22 Whitehouse PJ, Price DL, Clark AW, Coyle JT, DeLong MR: Alzheimer disease: evidence for selective loss of cholinergic neurons in the nucleus basalis. Ann Neurol 1981;10:122–126.
23 Whitehouse PJ, Struble RG, Clark AW, Price DL: Alzheimer disease: plaques, tangles, and the basal forebrain. Ann Neurol 1982;12:494.
24 Wilcock GK, Esiri MM, Bowen DM, Smith CC: Alzheimer's disease. Correlation of cortical choline acetyltransferase activity with the severity of dementia and histological abnormalities. J Neurol Sci 1982;57:407–417.
25 Davis KL, Thal LJ, Gamzu ER, Davis CS, Woolson RF, Gracon SI, Drachman DA, Schneider LS, Whitehouse PJ, Hoover TM, et al: A double-blind, placebo-controlled multicenter study of tacrine for Alzheimer's disease. The Tacrine Collaborative Study Group. N Engl J Med 1992;327:1253–1259.
26 Rogers SL, Friedhoff LT: The efficacy and safety of donepezil in patients with Alzheimer's disease: results of a US Multicentre, Randomized, Double-Blind, Placebo-Controlled Trial. The Donepezil Study Group. Dementia 1996;7:293–303.
27 Glenner GG, Wong CW: Alzheimer's disease: initial report of the purification and characterization of a novel cerebrovascular amyloid protein. Biochem Biophys Res Commun 1984;120:885–890.
28 Masters CL, Simms G, Weinman NA, Multhaup G, McDonald BL, Beyreuther K: Amyloid plaque core protein in Alzheimer disease and Down syndrome. Proc Natl Acad Sci USA 1985;82:4245–4249.
29 Cummings BJ, Cotman CW: Image analysis of β-amyloid load in Alzheimer's disease and relation to dementia severity. Lancet 1995;346:1524–1528.
30 Klunk WE, Engler H, Nordberg A, Wang Y, Blomqvist G, Holt DP, Bergström M, Savitcheva I, Huang GF, Estrada S, Ausén B, Debnath ML, Barletta J, Price JC, Sandell J, Lopresti BJ, Wall A, Koivisto P, Antoni G, Mathis CA, Långström B: Imaging brain amyloid in Alzheimer's disease with Pittsburgh Compound-B. Ann Neurol 2004;55: 306–319.
31 Clark CM, Schneider JA, Bedell BJ, Beach TG, Bilker WB, Mintun MA, Pontecorvo MJ, Hefti F, Carpenter AP, Flitter ML, Krautkramer MJ, Kung HF, Coleman RE, Doraiswamy PM, Fleisher AS, Sabbagh MN, Sadowsky CH, Reiman EP, Zehntner SP, Skovronsky DM, AV45-A07 Study Group: Use of florbetapir-PET for imaging β-amyloid pathology. JAMA 2011;305:275–283.
32 Gilman S, Koller M, Black RS, Jenkins L, Griffith SG, Fox NC, Eisner L, Kirby L, Rovira MB, Forette F, Orgogozo JM, AN1792(QS-21)-201 Study Team: Clinical effects of Aβ immunization (AN1792) in patients with AD in an interrupted trial. Neurology 2005;64:1553–1562.
33 Green RC, Schneider LS, Amato DA, Beelen AP, Wilcock G, Swabb EA, Zavitz KH, Tarenflurbil Phase 3 Study Group: Effect of tarenflurbil on cognitive decline and activities of daily living in patients with mild Alzheimer disease: a randomized controlled trial. JAMA 2009;302:2557–2564.
34 Katzman R, Terry R, DeTeresa R, Brown T, Davies P, Fuld P, Renbing X, Peck A: Clinical, pathological, and neurochemical changes in dementia: a subgroup with preserved mental status and numerous neocortical plaques. Ann Neurol 1988;23:138–144.
35 Crystal H, Dickson D, Fuld P, Masur D, Scott R, Mehler M, Masdeu J, Kawas C, Aronson M, Wolfson L: Clinico-pathologic studies in dementia: nondemented subjects with pathologically confirmed Alzheimer's disease. Neurology 1988;38:1682–1687.

36 Petersen RC, Stevens JC, Ganguli M, Tangalos EG, Cummings JL, DeKosky ST: Practice parameter: early detection of dementia: mild cognitive impairment (an evidence-based review). Report of the Quality Standards Subcommittee of the American Academy of Neurology. Neurology 2001; 56:133–1142.

37 Graham JE, Rockwood K, Beattie BL, Eastwood R, Gauthier S, Tuokko H, McDowell I: Prevalence and severity of cognitive impairment with and without dementia in an elderly population. Lancet 1997;349: 1793–1796.

38 Ditter SM, Mirra SS: Neuropathologic and clinical features of Parkinson's disease in Alzheimer's disease patients. Neurology 1987;37:754–760.

39 Kazee AM, Eskin TA, Lapham LW, Gabriel KR, McDaniel KD, Hamill RW: Clinicopathologic correlates in Alzheimer disease: assessment of clinical and pathologic diagnostic criteria. Alzheimer Dis Assoc Disord 1993;7:152–164.

40 Schneider JA, Arvanitakis Z, Bang W, Bennett DA: Mixed Brain pathologies account for most dementia cases in community-dwelling older persons. Neurology 2007;69:2197–2204.

41 Sonnen JA, Postupna N, Larson EB, Crane PK, Rose SE, Montine KS, Leverenz JB, Montine TJ: Pathological correlates of dementia in a longitudinal, population-based sample of aging. Ann Neurol 2007;62:406–413.

42 White L, Small BJ, Petrovitch H, Ross GW, Masaki K, Abbott RD, Hardman J, Davis D, Nelson J, Markesbery W: Recent clinical-pathologic research on the causes of dementia in later life: update from the Honolulu-Asia Aging Study. J Geriatr Psychiatry Neurol 2005;18:224–227.

43 Neuropathology Group of the Medical Research Council Cognitive Function and Aging Study (MRC CFAS): Pathologic correlates of late onset dementia in a multicentre, community-based population in England and Wales. Lancet 2001;357:169–175.

44 Nelson PT, Abner EL, Schmitt FA, Kryscio RJ, Jicha GA, Smith CD, Davis DG, Poduska JW, Patel E, Mendiondo MS, Markesbery WR: Modeling the association between 43 different clinical and pathological variables and the severity of cognitive impairment in a large autopsy cohort of elderly persons. Brain Pathol 2010;20:66–79.

45 O'Brien RJ, Resnick SM, Zonderman AB, Ferrucci L, Crain BJ, Pletnikova O, Rudow G, Iacono D, Riudavets MA, Driscoll I, Price DL, Martin LJ, Troncoso JC: Neuropathologic studies of the Baltimore Longitudinal Study of Aging (BLSA). J Alzheimers Dis 2009;18:665–675.

46 Thal DR, Capetillo-Zarate E, Del Tredici K, Braak H: The development of amyloid-β protein deposits in the aged brain. Sci Aging Knowledge Environ 2006:re1.

47 Thal DR, Rüb U, Schultz C, Sassin I, Ghebremedhin E, Del Tredici K, Braak E, Braak H: Sequence of Aβ-protein deposition in the human medial temporal lobe. J Neuropathol Exp Neurol 2000;59:733–748.

48 Braak H, Braak E: Neuropathological staging of Alzheimer-related changes. Acta Neuropathol (Berl) 1991;82:239–259.

49 Coomaraswamy J, Kilger E, Wölfing H, Schäfer C, Kaeser SA, Wegenast-Braun BM, Hefendehl JK, Wolburg H, Mazzella M, Ghiso J, Goedert M, Akiyama H, Garcia-Sierra F, Wolfer DP, Mathews PM, Jucker M: Modeling familial Danish dementia in mice supports the concept of the amyloid hypothesis of Alzheimer's disease. Proc Natl Acad Sci USA 2010;107:7969–7974.

50 Bennett DA, Schneider JA, Wilson RS, Bienias JL, Arnold SE: Neurofibrillary tangles mediate the association of amyloid load with clinical Alzheimer disease and level of cognitive function. Arch Neurol 2004;61:378–384.

51 Lo RY, Hubbard AE, Shaw LM, Trojanowski JQ, Petersen RC, Aisen PS, Weiner MW, Jagust WJ: Alzheimer's Disease Neuroimaging Initiative: Longitudinal change of biomarkers in cognitive decline. Arch Neurol 2011;68:1257–1266.

52 Jack CR Jr, Knopman DS, Jagust WJ, Shaw LM, Aisen PS, Weiner MW, Petersen RC, Trojanowski JQ: Hypothetical model of dynamic biomarkers of the Alzheimer's pathological cascade. Lancet Neurol 2010;9:119–112.

53 Braak H, Del Tredici K: The pathological process underlying Alzheimer's disease in individuals under thirty. Acta Neuropathol 2011;121:171–181.

54 Duyckaerts C: Tau pathology in children and young adults: can you still be unconditionally baptist? Acta Neuropathol 2011;121:145–147.

55 Mouton PR, Martin LJ, Calhoun ME, Dal Forno G, Price DL: Cognitive decline strongly correlates with cortical atrophy in Alzheimer's dementia. Neurol Aging 1998;19:371–377.

56 Berg L, McKeel DW Jr, Miller JP, Storandt M, Rubin EH, Morris JC, Baty J, Coats M, Norton J, Goate AM, Price JL, Gearing M, Mirra SS, Saunders AM: Clinicopathologic studies in cognitively healthy aging and Alzheimer's disease: relation of histologic markers to dementia severity, age, sex, and apolipoprotein E genotype. Arch Neurol 1998;55:326–335.

57 Arnold SE, Hyman BT, Flory J, Damasio AR, Van Hoesen GW: The topographical and neuroanatomical distribution of neurofibrillary tangles and neuritic plaques in the cerebral cortex of patients with Alzheimer's disease. Cereb Cortex 1991;1:103–116.

58 Arvanitakis Z, Leurgans SE, Wang Z, Wilson RS, Bennett DA, Schneider JA: Cerebral amyloid angiopathy pathology and cognitive domains in older persons. Ann Neurol 2011;69:320–327.

59 Thal DR, Griffin WS, Braak H: Parenchymal and vascular Aβ-deposition and its effects on the degeneration of neurons and cognition in Alzheimer's disease. J Cell Mol Med 2008;12:1848–1862.

60 New Diagnostic Criteria and Guidelines for Alzheimer's Disease: http://www.alz.org/research/diagnostic_criteria/#newreco.

61 Bennett DA, Schneider JA, Arvanitakis Z, Kelly JF, Aggarwal NT, Shah RC, Wilson RS: Neuropathology of older persons without cognitive impairment from two community-based studies. Neurology 2006;66:1837–1844.

62 Schmitt FA, Davis DG, Wekstein DR, Smith CD, Ashford JW, Markesbery WR: 'Preclinical' AD revisited: neuropathology of cognitively normal older adults. Neurology 2000;55:370–376.

63 Price JL, McKeel DW Jr, Buckles VD, Roe CM, Xiong C, Grundman M, Hansen LA, Petersen RC, Parisi JE, Dickson DW, Smith CD, Davis DG, Schmitt FA, Markesbery WR, Kaye J, Kurlan R, Hulette C, Kurland BF, Higdon R, Kukull W, Morris JC: Neuropathology of nondemented aging: presumptive evidence for preclinical Alzheimer disease. Neurobiol Aging 2009;30:1026–1036.

64 Hampel H, Shen Y, Walsh DM, Aisen P, Shaw LM, Zetterberg H, Trojanowski JQ, Blennow K: Biological markers of amyloid-β-related mechanisms in Alzheimer's disease. Exp Neurol 2010;223:334–346.

65 Mintun MA, Larossa GN, Sheline YI, Dence CS, Lee SY, Mach RH, Klunk WE, Mathis CA, DeKosky ST, Morris JC: ^{11}C-PIB in a nondemented population: potential antecedent marker of Alzheimer disease. Neurology 2006;67:446–452.

66 Villemagne V, PikeK, Fodero-Tavoletti M, Jones G, McLean C, Hinton F, O'Keefe G, Cappai R, Masters C, Rowe C: Age dependent prevalence of β-amyloid positive ^{11}C-PIB PET in healthy elderly subjects parallels neuropathology findings. J Nucl Med 2008;49(suppl 1):34P.

67 Sojkova J, Zhou Y, An Y, Kraut MA, Ferrucci L, Wong DF, Resnick SM: Longitudinal patterns of β-amyloid deposition in nondemented older adults. Arch Neurol 2011;68:644–649.

68 Stomrud E, Hansson O, Zetterberg H, Blennow K, Minthon L, Londos E: Correlation of longitudinal cerebrospinal fluid biomarkers with cognitive decline in healthy older adults. Arch Neurol 2010;67:217–223.

69 Villemagne VL, Pike KE, Darby D, Maruff P, Savage G, Ng S, Ackermann U, Cowie TF, Currie J, Chan SG, Jones G, Tochon-Danguy H, O'Keefe G, Masters CL, Rowe CC: Aβ deposits in older non-demented individuals with cognitive decline are indicative of preclinical Alzheimer's disease. Neuropsychologia 2008;46:1688–1697.

70 Pike KE, Savage G, Villemagne VL, Ng S, Moss SA, Maruff P, Mathis CA, Klunk WE, Masters CL, Rowe CC: Beta-amyloid imaging and memory in non-demented individuals: evidence for preclinical Alzheimer's disease. Brain 2007;130:2837–2844.

71 Mormino EC, Kluth JT, Madison CM, Rabinovici GD, Baker SL, Miller BL, Koeppe RA, Mathis CA, Weiner MW, Jagust WJ: Alzheimer's Disease Neuroimaging Initiative. Episodic memory loss is related to hippocampal-mediated β-amyloid deposition in elderly subjects. Brain 2009;132:1310–1323.

72 Rentz DM, Locascio JJ, Becker JA, Moran EK, Eng E, Buckner RL, Sperling RA, Johnson KA: Cognition, reserve, and amyloid deposition in normal aging. Ann Neurol 2010;67:353–364.

73 Roe CM, Fagan AM, Grant EA, Marcus DS, Benzinger TL, Mintun MA, Holtzman DM, Morris JC: Cerebrospinal fluid biomarkers, education, brain volume, and future cognition. Arch Neurol 2011;68:1145–1151.

74 Resnick SM, Sojkova J: Amyloid imaging and memory change for prediction of cognitive impairment. Alzheimers Res Ther 2011;3:3.

75 Barnes LL, Schneider JA, Boyle PA, Bienias JL, Bennett DA: Memory complaints are related to Alzheimer disease pathology in older persons. Neurology 2006;67:1581–1585.

76 Tucker AM, Stern Y: Cognitive reserve in aging. Curr Alzheimer Res 2011;8:354–360.

77 Snowdon DA, Greiner LH, Mortimer JA, Riley KP, Greiner PA, Markesbery WR: Brain infarction and the clinical expression of AD. The Nun Study. JAMA 1997;277:813–817.

78 Esiri MM, Nagy Z, Smith MZ, Barnetson L, Smith AD: Cerebrovascular disease and threshold for dementia in the early stages of Alzheimer's disease. Lancet 1999;354:919–920.

79 Schneider JA, Wilson RS, Bienias JL, Evans DA, Bennett DA: Cerebral infarctions and the likelihood of dementia from Alzheimer's disease pathology. Neurology 2004;62:1148–1156.

80 Peila R, Rodriguez BL, Launer LJ: Type 2 diabetes, APOE gene, and the risk for dementia and related pathologies: the Honolulu-Asia Aging Study. Diabetes 2002;51:1256–1262.

81 Wilson RS, Schneider JA, Arnold SE, Bienias JL, Bennett DA: Conscientiousness and the incidence of Alzheimer disease and mild cognitive impairment. Arch Gen Psychiatry 2007;64:1204–1212.
82 Bennett DA, Wilson RS, Schneider JA, Evans DA, Mendes de Leon CF, Arnold SE, Barnes LL, Bienias JL: Education modifies the relation of AD pathology to level of cognitive function in older persons. Neurology 2003;60:1909–1915.
83 Mormino EC, Brandel MG, Madison CM, Rabinovici GD, Marks S, Baker SL, Jagust WJ: Not quite PIB-positive, not quite PIB-negative: slight PIB elevations in elderly normal control subjects are biologically relevant. Neuroimage 2012;59: 1152–1160.
84 Wilson RS, Leurgans SE, Boyle PA, Schneider JA, Bennett DA: Neurodegenerative basis of cognitive decline in old age. Neurology 2010;75:1070–1078.
85 Mufson EJ, Chen EY, Cochran EJ, Beckett LA, Bennett DA, Kordower JH: Entorhinal cortex β-amyloid load in individuals with mild cognitive impairment. Exp Neurol 1999;158:469–490.
86 Mitchell TW, Mufson EJ, Schneider JA, Cochran EJ, Nissanov J, Han LY, Bienias JL, Lee VM, Trojanowski JQ, Bennett DA, Arnold SE: Parahippocampal tau pathology in healthy aging, mild cognitive impairment, and early Alzheimer's disease. Ann Neurol 2002;51:182–189.
87 Riley KP, Snowdon DA, Markesbery WR: Alzheimer's neurofibrillary pathology and the spectrum of cognitive function: findings from the Nun Study. Ann Neurol 2002;51:567–577.
88 Galvin JE, Powlishta KK, Wilkins K, McKeel DW Jr, Xiong C, Grant E, Storandt M, Morris JC: Predictors of preclinical Alzheimer disease and dementia: a clinicopathologic study. Arch Neurol 2005;62: 758–765.
89 Bennett DA, Schneider JA, Bienias JL, Evans DA, Wilson RS: Mild cognitive impairment is related to Alzheimer disease pathology and cerebral infarctions. Neurology 2005;64:834–841.
90 Markesbery WR, Schmitt FA, Kryscio RJ, Davis DG, Smith CD, Wekstein DR: Neuropathologic substrate of mild cognitive impairment. Arch Neurol 2006; 63:38–46.
91 Petersen RC, Parisi JE, Dickson DW, Johnson KA, Knopman DS, Boeve BF, Jicha GA, Ivnik RJ, Smith GE, Tangalos EG, Braak H, Kokmen E: Neuropathologic features of amnestic mild cognitive impairment. Arch Neurol 2006;63:665–672.
92 Schneider JA, Arvanitakis Z, Leurgans SE, Bennett DA: The neuropathology of probable Alzheimer disease and mild cognitive impairment. Ann Neurol 2009;66:200–208.
93 Matthews FE, Brayne C, Lowe J, McKeith I, Wharton SB, Ince P: Epidemiological pathology of dementia: attributable risks at death in the Medical Research Council Cognitive Function and Ageing Study. PLoS Med 2009;6:e1000180.
94 Petrovitch H, Ross GW, Steinhorn SC, et al: AD lesions and infarcts in demented and non-demented Japanese-American men. Ann Neurol 2005;57:98–103.
95 Troncoso JC, Zonderman AB, Resnick SM, Crain B, Pletnikova O, O'Brien RJ: Effect of infarcts on dementia in the Baltimore longitudinal study of aging. Ann Neurol 2008;64:168–176.
96 Schneider JA, Aggarwal NT, Barnes L, Boyle P, Bennett DA: The neuropathology of older persons with and without dementia from community versus clinic cohorts. J Alzheimers Dis 2009;18:691–701.
97 Arvanitakis Z, Wilson RS, Bienias JL, Evans DA, Bennett DA: Diabetes mellitus and risk of Alzheimer disease and decline in cognitive function. Arch Neurol 2004;61:661–666.
98 Arvanitakis Z, Schneider JA, Wilson RS, Li Y, Arnold SE, Wang Z, Bennett DA: Diabetes is related to cerebral infarction but not to AD pathology in older persons. Neurology 2006;67:1960–1965.
99 Armstrong RA: The pathogenesis of Alzheimer's disease: a reevaluation of the 'amyloid cascade hypothesis'. Int J Alzheimers Dis 2011:630865.
100 Dowling NM, Tomaszewski Farias S, Reed BR, Sonnen JA, Strauss ME, Schneider JA, Bennett DA, Mungas D: Neuropathological associates of multiple cognitive functions in two community-based cohorts of older adults. J Int Neuropsychol Soc 2010;22:1–13.
101 Holtzman DM, Morris JC, Goate AM: Alzheimer's disease: the challenge of the second century. Sci Transl Med 2011;3:77sr1.
102 Cleary JP, Walsh DM, Hofmeister JJ, Shankar GM, Kuskowski MA, Selkoe DJ, Ashe KH: Natural oligomers of the amyloid-β protein specifically disrupt cognitive function. Nat Neurosci 2005;8:79–84.
103 Lesné S, Koh MT, Kotilinek L, Kayed R, Glabe CG, Yang A, Gallagher M, Ashe KH: A specific amyloid-β protein assembly in the brain impairs memory. Nature 2006;440:352–357.
104 Steinerman JR, Irizarry M, Scarmeas N, Raju S, Brandt J, Albert M, Blacker D, Hyman B, Stern Y: Distinct pools of β-amyloid in Alzheimer disease-affected brain: a clinicopathologic study. Arch Neurol 2008;65:906–912.
105 Reed MN, Hofmeister JJ, Jungbauer L, Welzel AT, Yu C, Sherman MA, Lesné S, Ladu MJ, Walsh DM, Ashe KH, Cleary JP: Cognitive effects of cell-derived and synthetically derived Aβ oligomers. Neurobiol Aging 2011;32:1784–1794.

106 Shankar GM, Li S, Mehta TH, Garcia-Munoz A, Shepardson NE, Smith I, Brett FM, Farrell MA, Rowan MJ, Lemere CA, Regan CM, Walsh DM, Sabatini BL, Selkoe DJ: Amyloid-β protein dimers isolated directly from Alzheimer's brains impair synaptic plasticity and memory. Nat Med 2008;14: 837–842.

107 Buchhave P, Janciauskiene S, Zetterberg H, Blennow K, Minthon L, Hansson O: Elevated plasma levels of soluble CD40 in incipient Alzheimer's disease. Neurosci Lett 2009;450:56–59.

108 Perneczky R, Tsolakidou A, Arnold A, Diehl-Schmid J, Grimmer T, Förstl H, Kurz A, Alexopoulos P: CSF soluble amyloid precursor proteins in the diagnosis of incipient Alzheimer disease. Neurology 2011;77:35–38.

109 Gao CM, Yam AY, Wang X, Magdangal E, Salisbury C, Peretz D, Zuckermann RN, Connolly MD, Hansson O, Minthon L, Zetterberg H, Blennow K, Fedynyshyn JP, Allauzen S: $A\beta_{40}$ oligomers identified as a potential biomarker for the diagnosis of Alzheimer's disease. PLoS One 2010;5:e15725.

110 Jin M, Shepardson N, Yang T, Chen G, Walsh D, Selkoe DJ: Soluble amyloid β-protein dimers isolated from Alzheimer cortex directly induce tau hyperphosphorylation and neuritic degeneration. Proc Natl Acad Sci USA 2011;108:5819–5824.

111 Dodart JC, Bales KR, Gannon KS, Greene SJ, DeMattos RB, Mathis C, DeLong CA, Wu S, Wu X, Holtzman DM, Paul SM: Immunization reverses memory deficits without reducing brain Aβ burden in Alzheimer's disease model. Nat Neurosci 2002;5: 452–457.

112 Hardy J: The amyloid hypothesis for Alzheimer's disease: a critical reappraisal. J Neurochem 2009;110: 1129–1134.

113 Monsonego A, Imitola J, Petrovic S, Zota V, Nemirovsky A, Baron R, Fisher Y, Owens T, Weiner HL: Aβ-induced meningoencephalitis is IFN-γ-dependent and is associated with T-cell-dependent clearance of Aβ in a mouse model of Alzheimer's disease. Proc Natl Acad Sci USA 2006;103:5048–5053.

114 Kadokura A, Yamazaki T, Lemere CA, Takatama M, Okamoto K: Regional distribution of TDP-43 inclusions in Alzheimer disease brains: their relation to Alzheimer disease common pathology. Neuropathology 2009;29:566–573.

115 Cairns NJ, Neumann M, Bigio EH, Holm IE, Troost D, Hatanpaa KJ, Foong C, White CL 3rd, Schneider JA, Kretzschmar HA, Carter D, Taylor-Reinwald L, Paulsmeyer K, Strider J, Gitcho M, Goate AM, Morris JC, Mishra M, Kwong LK, Stieber A, Xu Y, Forman MS, Trojanowski JQ, Lee VM, Mackenzie IR: TDP-43 in familial and sporadic frontotemporal lobar degeneration with ubiquitin inclusions. Am J Pathol 2007;171:227–240.

116 Amador-Ortiz C, Lin WL, Ahmed Z, Personett D, Davies P, Duara R, Graff-Radford NR, Hutton ML, Dickson DW: TDP-43 immunoreactivity in hippocampal sclerosis and Alzheimer's disease. Ann Neurol 2007;61:435–445.

117 Uryu K, Nakashima-Yasuda H, Forman MS, Kwong LK, Clark CM, Grossman M, Miller BL, Kretzschmar HA, Lee VM, Trojanowski JQ, Neumann M: Concomitant TAR-DNA-binding protein-43 pathology is present in Alzheimer disease and corticobasal degeneration but not in other tauopathies. J Neuropathol Exp Neurol 2008;67:555–564.

118 Hu WT, Josephs KA, Knopman DS, Boeve BF, Dickson DW, Petersen RC, Parisi JE: Temporal lobar predominance of TDP-43 neuronal cytoplasmic inclusions in Alzheimer disease. Acta Neuropathol 2008;116:215–220.

119 Josephs KA, Whitwell JL, Knopman DS, Hu WT, Stroh DA, Baker M, Rademakers R, Boeve BF, Parisi JE, Smith GE, Ivnik RJ, Petersen RC, Jack CR Jr, Dickson DW: Abnormal TDP-43 immunoreactivity in AD modifies clinicopathologic and radiologic phenotype. Neurology 2008;70:1850–1857.

120 Dubois B, Feldman HH, Jacova C, Dekosky ST, Barberger-Gateau P, Cummings J, Delacourte A, Galasko D, Gauthier S, Jicha G, Meguro K, O'Brien J, Pasquier F, Robert P, Rossor M, Salloway S, Stern Y, Visser PJ, Scheltens P: Research criteria for the diagnosis of Alzheimer's disease: revising the NINCDS-ADRDA criteria. Lancet Neurol 2007;6: 734–746.

121 Tremblay C, St-Amour I, Schneider J, Bennett DA, Calon F: Accumulation of transactive response DNA binding protein 43 in mild cognitive impairment and Alzheimer disease. J Neuropathol Exp Neurol 2011;70:788–798.

122 McKhann GM, Knopman DS, Chertkow H, Hyman BT, Jack CR Jr, Kawas CH, Klunk WE, Koroshetz WJ, Manly JJ, Mayeux R, Mohs RC, Morris JC, Rossor MN, Scheltens P, Carrillo MC, Thies B, Weintraub S, Phelps CH: The diagnosis of dementia due to Alzheimer's disease: recommendations from the National Institute on Aging-Alzheimer's Association workgroups on diagnostic guidelines for Alzheimer's disease. Alzheimers Dement 2011;7: 263–269.

123 Albert MS, DeKosky ST, Dickson D, Dubois B, Feldman HH, Fox NC, Gamst A, Holtzman DM, Jagust WJ, Petersen RC, Snyder PJ, Carrillo MC, Thies B, Phelps CH: The diagnosis of mild cognitive impairment due to Alzheimer's disease: recommendations from the National Institute on Aging-Alzheimer's Association workgroups on diagnostic guidelines for Alzheimer's disease. Alzheimers Dement 2011;7:270–279.

124 Sperling RA, Aisen PS, Beckett LA, Bennett DA, Craft S, Fagan AM, Iwatsubo T, Jack CR Jr, Kaye J, Montine TJ, Park DC, Reiman EM, Rowe CC, Siemers E, Stern Y, Yaffe K, Carrillo MC, Thies B, Morrison-Bogorad M, Wagster MV, Phelps CH: Toward defining the preclinical stages of Alzheimer's disease: recommendations from the National Institute on Aging-Alzheimer's Association workgroups on diagnostic guidelines for Alzheimer's disease. Alzheimers Dement 2011;7:280–292.

125 Jack CR Jr, Albert MS, Knopman DS, McKhann GM, Sperling RA, Carrillo MC, Thies B, Phelps CH: Introduction to the recommendations from the National Institute on Aging-Alzheimer's Association workgroups on diagnostic guidelines for Alzheimer's disease. Alzheimers Dement 2011;7:257–262.

126 Ikonomovic MD, Klunk WE, Abrahamson EE, Mathis CA, Price JC, Tsopelas ND, Lopresti BJ, Ziolko S, Bi W, Paljug WR, Debnath ML, Hope CE, Isanski BA, Hamilton RL, DeKosky ST: Postmortem correlates of in vivo PiB-PET amyloid imaging in a typical case of Alzheimer's disease. Brain 2008;131:1630–1645.

127 Leinonen V, Alafuzoff I, Aalto S, Suotunen T, Savolainen S, Någren K, Tapiola T, Pirttilä T, Rinne J, Jääskeläinen JE, Soininen H, Rinne JO: Assessment of β-amyloid in a frontal cortical brain biopsy specimen and by positron emission tomography with carbon 11-labeled Pittsburgh Compound B. Arch Neurol 2008;65:1304–1309.

128 Weigand SD, Vemuri P, Wiste HJ, Senjem ML, Pankratz VS, Aisen PS, Weiner MW, Petersen RC, Shaw LM, Trojanowski JQ, Knopman DS, Jack CR Jr: Alzheimer's Disease Neuroimaging Initiative. Transforming cerebrospinal fluid $A\beta_{42}$ measures into calculated Pittsburgh Compound B units of brain Aβ amyloid. Alzheimers Dement 2011;7: 133–141.

129 De Souza LC, Chupin M, Lamari F, Jardel C, Leclercq D, Colliot O, Lehéricy S, Dubois B, Sarazin M: CSF tau markers are correlated with hippocampal volume in Alzheimer's disease. Neurobiol Aging 2011 (E-pub ahead of print).

130 Whitwell JL, Jack CR Jr, Przybelski SA, Parisi JE, Senjem ML, Boeve BF, Knopman DS, Petersen RC, Dickson DW, Josephs KA: Temporoparietal atrophy: a marker of AD pathology independent of clinical diagnosis. Neurobiol Aging 2011;32: 1531–1541.

131 Jagust WJ, Zheng L, Harvey DJ, Mack WJ, Vinters HV, Weiner MW, Ellis WG, Zarow C, Mungas D, Reed BR, Kramer JH, Schuff N, DeCarli C, Chui HC: Neuropathological basis of magnetic resonance images in aging and dementia. Ann Neurol 2008;63:72–80.

132 Mosconi L, Pupi A, De Cristofaro MT, Fayyaz M, Sorbi S, Herholz K: Functional interactions of the entorhinal cortex: an ^{18}F-FDG-PET study on normal aging and Alzheimer's disease. J Nucl Med 2004;45:382–392.

133 Jack CR, Lowe VJ, Senjem ML, et al: ^{11}C-PiB and structural MRI provide complementary information in imaging of Alzheimer's disease and amnestic mild cognitive impairment. Brain 2008;131: 665–680.

Assoc. Prof. Julie A. Schneider, MD, MS
Department of Pathology (Neuropathology) and Neurological Sciences
Rush Alzheimer's Disease Center, Rush University Medical Center, Armour Academic Center
600 South Paulina Street, Suite 1022F, Chicago, IL 60612 (USA)
Tel. +1 312 942 2360, E-Mail julie_a_schneider@rush.edu

Hampel H, Carrillo MC (eds): Alzheimer's Disease – Modernizing Concept, Biological Diagnosis and Therapy.
Adv Biol Psychiatry. Basel, Karger, 2012, vol 28, pp 71–79

Diagnostic Tools: Fluid Biomarkers for Alzheimer's Disease

Henrik Zetterberg[a] · Harald Hampel[b] · Kaj Blennow[a]

[a]Clinical Neurochemistry Laboratory, Institute of Neuroscience and Physiology, Department of Psychiatry and Neurochemistry, The Sahlgrenska Academy at University of Gothenburg, Mölndal, Sweden, and [b]Department of Psychiatry, Psychosomatic Medicine and Psychotherapy, Goethe University, Frankfurt a.M., Germany

Abstract

Intense multidisciplinary research has provided detailed knowledge on the molecular mechanisms underlying Alzheimer's disease (AD). This knowledge has been translated into new therapeutic strategies that target early pathological processes. New diagnostic criteria for early AD have been proposed and rely upon biomarker evidence of AD pathology in conjunction with mild clinical symptoms. In this chapter, we present the core cerebrospinal fluid biomarkers for AD, namely total tau (T-tau), phosphorylated tau (P-tau) and the 42-amino-acid form of amyloid β ($A\beta_{42}$), and discuss how they might fit into new diagnostic algorithms for pre-dementia AD. We also discuss new biomarker candidates as well as validation and standardization issues.

Currently employed clinical criteria for Alzheimer's disease (AD), established more than 25 years ago, require that the patient is demented before a diagnosis of AD can be made [1]. However, today we know that this represents an advanced stage of AD and that causative molecular processes start many years, if not decades, earlier. Disease-modifying drugs that attack primary pathogenic processes underlying AD will probably be most effective if treatment is initiated before plaque and tangle load and neurodegeneration have become too widespread and severe. This has necessitated revision of the clinical criteria so that they also recognize pre-dementia stages of the disease.

Thanks to brain imaging and cerebrospinal fluid (CSF) analyses, it has become possible to measure AD pathology in living patients and relate biomarker changes to clinical presentation and disease course [2–4]. New biomarker-based research criteria have been proposed that capture both prodromal and more advanced dementia stages of AD in the same diagnostic framework [5]. Further, a new lexicon that elaborates on the definition of AD and how pre-dementia AD could be diagnosed using biomarkers has been published [6]. Along similar lines, new clinical criteria incorporating biomarkers were recently proposed [7]. This chapter discusses fluid biomarkers for AD and their possible role in new diagnostic criteria.

Biomarker Source

Because of its proximity to the brain parenchyma and interchange with brain extracellular fluid, the composition of CSF can provide insights into the brain chemistry and cellular processes, and because it is also separated from the blood by barriers that restrict molecular and cellular passage, it can also serve as a 'model' of an isolated and protected tissue sharing properties with brain [8]. Conversely, peripheral body fluids are separated from the brain by the blood-brain barrier (BBB), which substantially limits their use as biomarker sources. These unique characteristics of the CSF compartment, along with the safety of CSF sampling [9], have contributed to reaching an informative view of the association of AD pathology with clinical symptoms of the disease. While CSF studies must be interpreted carefully, since the composition of CSF reflects a variable mix of meningeal and brain components, this body fluid has proven informative and provided insight into otherwise inaccessible processes in AD. For example, longitudinal CSF biomarker studies indicate that amyloid pathology precedes tau pathology and cognitive decline by several years [9–12], which is in line with data from animal models [13, 14].

It has been much more difficult to find reliable biomarkers for AD in peripheral blood, most likely due to the low abundance of brain-derived proteins. Promising pilot studies on plasma exist but they are all in need of replication and far from being established [15–17]. Groundbreaking new technology that allows for single molecule detection in biological fluids may change this statement within the next few years [18], but today no validated diagnostically useful peripheral blood biomarker for AD exists.

Biomarkers for AD Pathology

CSF levels of total tau (T-tau) reflect cortical axonal degeneration, phospho-tau (P-tau) is related to tangle pathology and the 42-amino-acid isoform of amyloid β ($A\beta_{42}$) reflects brain amyloid pathology [2], i.e. the core elements of AD pathology. Together, these biomarkers are 80–95% sensitive and specific for AD both in the dementia phase of the disorder and in prodromal disease stages [2]. The well-established age-related increase in pre-clinical AD neuropathology [19] is also reflected in the CSF biomarkers leading to lower diagnostic accuracy in patients older than 80 years of age compared with younger subjects [20]. In essence, the biomarkers are most useful to detect AD in younger subjects, and to rule out AD in older subjects. A more intense disease process is reflected by more aberrant biomarker concentrations and highly abnormal values predict rapid cognitive decline [21]. Successful disease-modifying treatment of AD should reduce CSF tau protein concentrations [22], while it is much more difficult to predict the response of Aβ_{42} due to multiple factors (production, clearance, oligomerization and aggregation) that influence steady-state concentrations of the peptide in CSF.

Biomarkers to Identify Pathological Processes Not Typical of AD

Apart from positive biomarkers for AD, a number of CSF analyses may be useful to exclude other causes of AD-like cognitive symptoms.

CSF Biomarkers for BBB Dysfunction
The best-established biomarker for the integrity of the BBB is the ratio of the albumin concentration in CSF to serum (the CSF/serum albumin ratio). Strictly speaking, the CSF/serum albumin ratio is a direct measure of the blood-CSF barrier (BCB) [23]. However, leaking blood vessels in the brain will result in higher CSF/serum albumin ratio through release of albumin to the brain interstitial fluid which communicates freely with the CSF. Typically, the CSF/serum albumin ratio is normal in patients with pure AD [24], whereas patients with vascular dementia generally present with elevated albumin ratio [25]. The same finding is often present in Lyme disease (neuroborreliosis), where one also may find increased numbers of CSF monocytes and signs of immunoglobulin production within the CNS [26]. Recently, secretory Ca^{2+}-dependent phospholipase A_2 ($sPLA_2$) activity in CSF was described as a sensitive measure of BCB permeability [27], detecting subtle BCB dysfunction in around 50% of AD patients [28].

Biomarkers for Neuroinflammation
Several standard CSF examinations may detect neuroinflammatory activity, including white blood cell count and general signs of IgG or IgM production within the CNS. These are generally negative in AD and other primary neurodegenerative diseases [29]. Distinct positive results speak against pure AD and should motivate further investigation of the patient to exclude neuroborreliosis, multiple sclerosis and other neuroinflammatory conditions that may contribute to the cognitive symptoms [30]. However, dysregulation within complex immunological and inflammatory pathways has been suggested to be a potentially crucial factor in the pathophysiology of AD [31]. Nonetheless, no diagnostically useful inflammation-based biomarker has been validated so far.

Biomarkers for Subcortical Axonal Degeneration
The best-established CSF biomarker for subcortical axonal degeneration/damage is neurofilament light protein (NFL). This type of axonal degeneration is frequently

seen in vascular dementia [32–34], frontotemporal dementia [35] and a number of inflammatory conditions, including multiple sclerosis [36] and HIV-associated dementia [37]. Elevated CSF NFL levels indicate subcortical disease and may help in differentiating pure AD from the conditions listed above. Combined T-tau and NFL increases suggest mixed forms of AD and subcortical cerebrovascular disease [38], which is very common in patients older than 80 years.

Peripheral Blood Markers of Metabolic Changes That May Impair Cognition
A number of standard clinical chemistry tests should be considered in patients who seek medical advice because of cognitive symptoms. These include thyroid hormone levels, markers of folic acid and cobalamin status, Ca^{2+} and Hb, since abnormal concentrations of these components often cause or associate with varying degrees of cognitive symptoms and represent conditions that are often amenable to treatment.

A Diagnostic Flow-Chart

Figure 1 presents a diagnostic flow-chart of how CSF biomarkers and blood tests may be employed to aid in the diagnostic work-up of patients with memory problems. The clinical examination will rapidly identify a number of patients with cognitive problems that are clearly related to diseases other than AD. Patients with clinical symptoms typical of AD may be investigated further to establish the diagnosis using biomarker evidence (CSF and/or neuroimaging) of AD pathology as an essential item. If biomarker evidence of AD pathology is lacking, the patient may suffer from other diseases that impair cognition or possibly have a benign condition with low risk of progression to dementia.

Standardization of CSF Biomarkers for AD

If CSF biomarkers (T-tau, P-tau and $A\beta_{42}$) are to be implemented on a global level and in clinical practice as part of diagnostic criteria for AD, several standardization issues need to be solved. Table 1 details these and provides recommendations on how we should move forward. In particular, we need to establish gold standard techniques for biomarker measurements and certified reference materials against which kit vendors can calibrate their assays. Standard operating procedures for the lumbar puncture and sample handling prior to analysis have already been published [2] and a global quality control program for CSF AD biomarker analyses is up and running [39]. Nevertheless, also in the absence of internationally accepted gold standard techniques and calibrators, CSF biomarkers may be used in the clinic if laboratories that know their longitudinal stability and can relate their biomarker measurements to diagnostic performance in clinical studies execute the analyses.

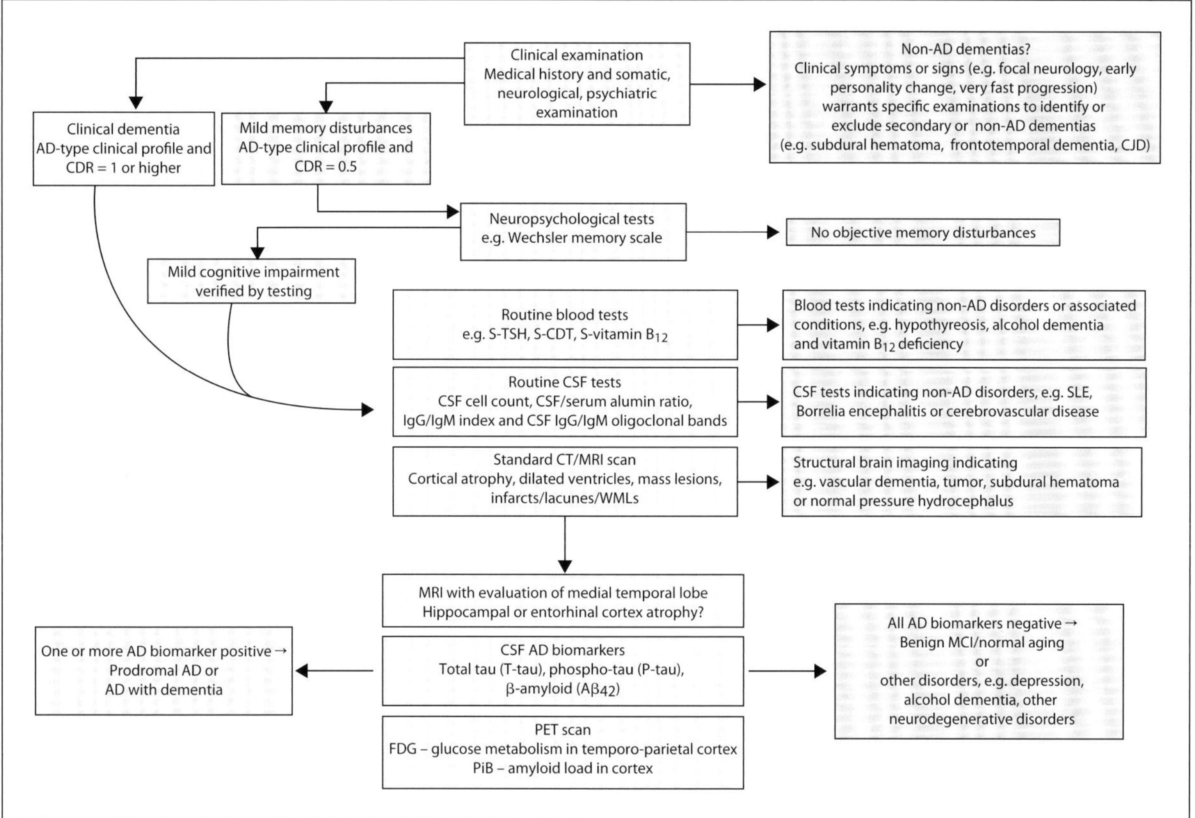

Fig. 1. The figure shows the central position of a thorough clinical examination of the patient in the diagnostic scheme. The result of this examination will guide further investigations in which different standard methods and evaluations to exclude somatic and non-neurodegenerative causes of cognitive impairment are central. When non-AD causes of cognitive problems have been excluded, biomarkers for AD pathology become important. Positive results on neuroimaging and/or CSF biomarkers for AD will lead to a diagnosis of prodromal AD or AD with dementia depending on disease severity. Negative biomarker results will exclude a diagnosis of AD. In our mind, this reasoning will prohibit that all kinds of patients end up with an AD diagnosis if no other cause of their symptoms can be found; a malpractice that is not uncommon today.

Conclusions

Numerous studies have shown that combined analysis of the core CSF biomarkers T-tau, P-tau and $A\beta_{42}$ can be used to reliably diagnose patients with AD and identify prodromal AD in patients with MCI. Standardization work of these biomarkers is moving forward and it should be possible to develop assays with analytical reproducibility and longitudinal stability similar to what we have for example for troponin T within the next few years. However, we also need additional biomarkers for other facets of the AD process, as well as for specific molecular changes in other neurodegenerative diseases. Using CSF biomarkers, it is still sometimes difficult to

Table 1. Standardization of CSF biomarkers for AD

	Problems	Recommendations	Comment	Status
Pre-analytical procedures	• Different procedures for LP causing variation due to for example adsorption of Aβ to test tubes or diurnal variation • Inadequate procedures for CSF sample shipment or storage	• Standardized procedures for LP • Standardized procedures for CSF sample handling, shipment and storage	• Standardized procedures for LP procedures and sample shipment should be communicated to clinicians • Standardized procedures for CSF sample handling and storage should be implemented in laboratories	• Flow-chart with recommendations for LP and CSF sample processing published
Laboratory and analytical procedures	• Laboratory instruments not regularly maintained, equipment such as pipettes not regularly calibrated • Recommended protocol for assays not followed • Assays performed without internal QC samples or specified run acceptance criteria	• Recommendations for maintenance and QC of instruments and equipment • Recommendations for standardized procedures for how to set up and run an assay	• Regular maintenance and calibration of instrumentation and equipment necessary • Assays should be performed by specially trained lab technicians • Controlled procedures for assay reagents and calibrators, and internal QC samples with specified run acceptance criteria should be applied	• Basic instructions available in kit inserts • Checklists for analysis of CSF biomarkers by ELISA and the MesoScale and Luminex techniques available, see: www.neurochem.gu.se/theAlzAssQCprogram
QC of assays	• Quality and purity of reagents in assays may vary • Procedures for production of assay kits may be suboptimal • New batches of kits are not compared to previous batches by analyses of CSF samples • Within-assay CV for individual batches, and CV between batches not specified by kit vendors	• QC of each component in an analytical kit • Highly controlled procedures for assay production with control of within-plate, between-plate and between-batch variation • Within-assay and between-batch CVs should be specified by kit vendors (e.g. in kit insert)	• Between-batch variation can be minimized by analyzing a set of CSF samples on several plates from a new batch and compare results with the previous assay batch	• Ongoing by kit vendors

Table 1. Continued

	Problems	Recommendations	Comment	Status
Standardiz-ation of CSF biomarkers within and between laboratories	• CSF biomarker levels vary between clinical laboratories • CSF biomarker levels vary over time	• International QC program • QC samples (aliquots of pooled CSF samples) are sent out to participating laboratories 3 times per year. Data with comparisons between labs are reported • An aliquot of the same QC sample is analyzed by the participating laboratories 3 times per year, to allow control of longitudinal variations	• Between-laboratory variations are due to differences in laboratory and analytical procedures, or to batch-to-batch variation for the analytical kits	• The Alzheimer's Association QC (QC) program is ongoing since 2010, see: www.neurochem.gu.se/theAlzAssQCprogram
Standardiz-ation between assay formats	• Differences in CSF biomarker levels between different assay formats	• International reference material for standardization of CSF biomarker levels between assay formats	• Standardization of biomarker levels between assays necessary to introduce uniform cut-off levels for CSF biomarkers	• Initiated

CV = Coefficient of variation; LP = lumbar puncture; QC = quality control.

differentiate AD from frontotemporal dementia and Lewy body dementia due to overlap in tau, amyloid and synuclein pathologies, which is reflected in the biomarker results. Biomarker areas of intense research include Aβ oligomers, synaptic proteins and markers of microglial activation. New single molecule ELISA techniques have also revived the search for blood-based biomarkers for neurodegenerative processes. Beside their emerging role in clinical practice for early and differential diagnosis of AD, fluid biomarkers play an increasingly important role in clinical therapeutic AD trials, where they can help to improve study design, enrich patient groups with 'true' AD cases and give support to clinical effect measurements [40].

Acknowledgements

Work in the authors' laboratories is supported by the Swedish Research Council and Alzheimer's Association.

References

1 McKhann G, Drachman D, Folstein M, Katzman R, Price D, Stadlan EM: Clinical diagnosis of Alzheimer's disease: report of the NINCDS-ADRDA Work Group under the auspices of Department of Health and Human Services Task Force on Alzheimer's Disease. Neurology 1984;34:939–944.

2 Blennow K, Hampel H, Weiner M, Zetterberg H: Cerebrospinal fluid and plasma biomarkers in Alzheimer disease. Nat Rev Neurol 2010;6:131–144.

3 Frisoni GB, Fox NC, Jack CR Jr, Scheltens P, Thompson PM: The clinical use of structural MRI in Alzheimer disease. Nat Rev Neurol 2010;6:67–77.

4 Nordberg A, Rinne JO, Kadir A, Langstrom B: The use of PET in Alzheimer disease. Nat Rev Neurol 2010;6:78–87.

5 Dubois B, Feldman HH, Jacova C, Dekosky ST, Barberger-Gateau P, Cummings J, Delacourte A, Galasko D, Gauthier S, Jicha G, et al: Research criteria for the diagnosis of Alzheimer's disease: revising the NINCDS-ADRDA criteria. Lancet Neurol 2007;6:734–746.

6 Dubois B, Feldman HH, Jacova C, Cummings JL, Dekosky ST, Barberger-Gateau P, Delacourte A, Frisoni G, Fox NC, Galasko D, et al: Revising the definition of Alzheimer's disease: a new lexicon. Lancet Neurol 2010;9:1118–1127.

7 http://www.alzheimersanddementia.org/content/ncg.

8 Engelhardt B, Sorokin L: The blood-brain and the blood-cerebrospinal fluid barriers: function and dysfunction. Semin Immunopathol 2009;31:497–511.

9 Gustafson DR, Skoog I, Rosengren L, Zetterberg H, Blennow K: Cerebrospinal fluid β-amyloid 1-42 concentration may predict cognitive decline in older women. J Neurol Neurosurg Psychiatry 2007;78:461–464.

10 Skoog I, Davidsson P, Aevarsson O, Vanderstichele H, Vanmechelen E, Blennow K: Cerebrospinal fluid β-amyloid 42 is reduced before the onset of sporadic dementia: a population-based study in 85-year-olds. Dement Geriatr Cogn Disord 2003;15:169–176.

11 Stomrud E, Hansson O, Blennow K, Minthon L, Londos E: Cerebrospinal fluid biomarkers predict decline in subjective cognitive function over 3 years in healthy elderly. Dement Geriatr Cogn Disord 2007;24:118–124.

12 Moonis M, Swearer JM, Dayaw MP, St George-Hyslop P, Rogaeva E, Kawarai T, Pollen DA: Familial Alzheimer disease: decreases in CSF $A\beta_{42}$ levels precede cognitive decline. Neurology 2005;65:323–325.

13 Gotz J, Chen F, van Dorpe J, Nitsch RM: Formation of neurofibrillary tangles in P301l tau transgenic mice induced by $A\beta_{42}$ fibrils. Science 2001;293:1491–1495.

14 Lewis J, Dickson DW, Lin WL, Chisholm L, Corral A, Jones G, Yen SH, Sahara N, Skipper L, Yager D, et al: Enhanced neurofibrillary degeneration in transgenic mice expressing mutant tau and APP. Science 2001;293:1487–1491.

15 Ray S, Britschgi M, Herbert C, Takeda-Uchimura Y, Boxer A, Blennow K, Friedman LF, Galasko DR, Jutel M, Karydas A, et al: Classification and prediction of clinical Alzheimer's diagnosis based on plasma signaling proteins. Nat Med 2007;13:1359–1362.

16 Hye A, Lynham S, Thambisetty M, Causevic M, Campbell J, Byers HL, Hooper C, Rijsdijk F, Tabrizi SJ, Banner S, et al: Proteome-based plasma biomarkers for Alzheimer's disease. Brain 2006;129:3042–3050.

17 Soares HD, Chen Y, Sabbagh M, Roher A, Schrijvers E, Breteler M: Identifying early markers of Alzheimer's disease using quantitative multiplex proteomic immunoassay panels. Ann NY Acad Sci 2009;1180:56–67.

18 Rissin DM, Kan CW, Campbell TG, Howes SC, Fournier DR, Song L, Piech T, Patel PP, Chang L, Rivnak AJ, et al: Single-molecule enzyme-linked immunosorbent assay detects serum proteins at subfemtomolar concentrations. Nat Biotechnol 2010;28:595–599.

19 Savva GM, Wharton SB, Ince PG, Forster G, Matthews FE, Brayne C: Age, neuropathology, and dementia. N Engl J Med 2009;360:2302–2309.

20 Mattsson N, Rosén E, Hansson O, Andreasen N, Parnetti L, Jonsson M, Herukka SK, van der Flier WM, Blankenstein MA, Ewers M, Rich K, Kaiser E, Verbeek MM, Rikkert MO, Tsolaki M, Mulugeta E, Aarsland D, Visser PJ, Schröder J, Marcusson J, de Leon M, Hampel H, Scheltens P, Wallin A, Eriksdotter Jönhagen M, Minthon L, Winblad B, Blennow K, Zetterberg H: Age and diagnostic performance of Alzheimer's disease CSF biomarkers. Neurology 2012, in press.

21 Wallin AK, Blennow K, Zetterberg H, Londos E, Minthon L, Hansson O: CSF biomarkers predict a more malignant outcome in Alzheimer disease. Neurology 2010;74:1531–1537.

22 Zetterberg H, Mattsson N, Blennow K, Olsson B: Use of theragnostic markers to select drugs for phase II/III trials for Alzheimer disease. Alzheimers Res Ther 2010;2:32.

23 Reiber H, Peter JB: Cerebrospinal fluid analysis: disease-related data patterns and evaluation programs. J Neurol Sci 2001;184:101–122.
24 Blennow K, Wallin A, Fredman P, Karlsson I, Gottfries CG, Svennerholm L: Blood-brain barrier disturbance in patients with Alzheimer's disease is related to vascular factors. Acta Neurol Scand 1990;81:323–326.
25 Wallin A, Blennow K, Rosengren L: Cerebrospinal fluid markers of pathogenetic processes in vascular dementia, with special reference to the subcortical subtype. Alzheimer Dis Assoc Disord 1999;13(suppl 3): S102–S105.
26 Tumani H, Nolker G, Reiber H: Relevance of cerebrospinal fluid variables for early diagnosis of neuroborreliosis. Neurology 1995;45:1663–1670.
27 Chalbot S, Zetterberg H, Blennow K, Fladby T, Grundke-Iqbal I, Iqbal K: Cerebrospinal fluid secretory Ca^{2+}-dependent phospholipase A_2 activity: a biomarker of blood-cerebrospinal fluid barrier permeability. Neurosci Lett 2010;478:179–183.
28 Chalbot S, Zetterberg H, Blennow K, Fladby T, Andreasen N, Grundke-Iqbal I, Iqbal K: Blood-cerebrospinal fluid barrier permeability in Alzheimer's disease. J Alzheimers Dis 2011;25:505–515.
29 Blennow K, Wallin A, Fredman P, Gottfries CG, Karlsson I, Svennerholm L: Intrathecal synthesis of immunoglobulins in patients with Alzheimer's disease. Eur Neuropsychopharmacol 1990;1:79–81.
30 Andreasen N, Blennow K, Zetterberg H: Neuroinflammation screening in immunotherapy trials against Alzheimer's disease. Int J Alzheimers Dis 2010:638379.
31 Herrup K: Reimagining Alzheimer's disease – an age-based hypothesis. J Neurosci 2010;30:16755–16762.
32 Rosengren LE, Karlsson JE, Sjogren M, Blennow K, Wallin A: Neurofilament protein levels in CSF are increased in dementia. Neurology 1999;52: 1090–1093.
33 Agren-Wilsson A, Lekman A, Sjoberg W, Rosengren L, Blennow K, Bergenheim AT, Malm J: CSF biomarkers in the evaluation of idiopathic normal pressure hydrocephalus. Acta Neurol Scand 2007; 116:333–339.
34 Wallin A, Sjogren M: Cerebrospinal fluid cytoskeleton proteins in patients with subcortical whitematter dementia. Mech Ageing Dev 2001;122: 1937–1949.
35 De Jong D, Jansen RW, Pijnenburg YA, van Geel WJ, Borm GF, Kremer HP, Verbeek MM: CSF neurofilament proteins in the differential diagnosis of dementia. J Neurol Neurosurg Psychiatry 2007;78: 936–938.
36 Norgren N, Sundstrom P, Svenningsson A, Rosengren L, Stigbrand T, Gunnarsson M: Neurofilament and glial fibrillary acidic protein in multiple sclerosis. Neurology 2004;63:1586–1590.
37 Gisslen M, Hagberg L, Brew BJ, Cinque P, Price RW, Rosengren L: Elevated cerebrospinal fluid neurofilament light protein concentrations predict the development of AIDS dementia complex. J Infect Dis 2007;195:1774–1778.
38 Bjerke M, Andreasson U, Rolstad S, Nordlund A, Lind K, Zetterberg H, Edman A, Blennow K, Wallin A: Subcortical vascular dementia biomarker pattern in mild cognitive impairment. Dement Geriatr Cogn Disord 2009;28:348–356.
39 Mattsson N, Blennow K, Zetterberg H: Interlaboratory variations in CSF biomarkers for Alzheimer's disease: united we stand, divided we fall. Clin Chem Lab Med 2010;48:603–607.
40 Hampel H, Frank R, Broich K, Teipel SJ, Katz RG, Hardy J, Herholz K, Bokde AL, Jessen F, Hoessler YC, et al: Biomarkers for Alzheimer's disease: academic, industry and regulatory perspectives. Nat Rev Drug Discov 2010;9:560–574.

Henrik Zetterberg, MD, PhD
Clinical Neurochemistry Laboratory, Institute of Neuroscience and Physiology
Department of Psychiatry and Neurochemistry, The Sahlgrenska Academy at University of Gothenburg
Sahlgrenska University Hospital, SE–431 80 Mölndal (Sweden)
Tel. +46 31 3430142, E-Mail henrik.zetterberg@gu.se

Hampel H, Carrillo MC (eds): Alzheimer's Disease – Modernizing Concept, Biological Diagnosis and Therapy.
Adv Biol Psychiatry. Basel, Karger, 2012, vol 28, pp 80–114

MRI- and PET-Based Imaging Markers for the Diagnosis of Alzheimer's Disease

Stefan Teipel[a] · Reisa A. Sperling[b] · Pawel Skudlarski[c] · Clifford Jack[d] · Harald Hampel[e] · Andreas Fellgiebel[f] · Karl Herholz[g]

[a]Department of Psychiatry, University of Rostock and DZNE, German Center for Neurodegenerative Diseases, Rostock, Germany; [b]Center for Alzheimer Research and Treatment, Department of Neurology, Brigham and Women's Hospital, Massachusetts General Hospital, Harvard Medical School, Boston, Mass.; [c]Olin Neuropsychiatry Research Center, Hartford Hospital/IOL, Hartford, Conn.; [d]Department of Radiology, Mayo Clinic, Rochester, Minn., USA; [e]Department of Psychiatry, Psychosomatic Medicine and Psychotherapy, Goethe University, Frankfurt a.M.; [f]Department of Psychiatry, University of Mainz, Mainz, Germany, and [g]University of Manchester, Manchester, UK

Abstract

Imaging markers of early neurodegeneration play an important role for the definition of predementia and preclinical stages of Alzheimer's disease according to the newly proposed diagnostic consensus criteria. Markers of regional and global brain atrophy in MRI and the detection of cortical metabolic decline and cerebral amyloid deposition using PET are the best established imaging markers for prodromal and clinical Alzheimer's disease to date. Detection of structural and functional cortical disconnection using functional MRI and diffusion tensor imaging add to the diagnostic sensitivity and specificity. Important areas of future research are the application of imaging markers in large multicenter studies, and their implementation in radiological expert systems for diagnosis. Additionally, we need to consider the effect of these new markers on care for patients and counseling of at-risk subjects.

The revised NINCDS-ADRDA criteria for the diagnosis of Alzheimer's disease (AD) [1] propose the use of imaging-derived biomarkers [2, 3] to increase the level of certainty of the diagnosis of dementia of the AD type (clinically highly probable AD) and to identify AD in predementia or even preclinical stages of the disease (http://www.alz.org/research/diagnostic_criteria). The new diagnostic entities of predementia or preclinical AD are intended to serve the early diagnosis in clinical diagnostic studies and the enrichment and stratification of samples for testing effects of potential disease modification in clinical trials of AD [4, 5].

The following three imaging modalities have widely been employed as diagnostic markers or as secondary endpoints in clinical studies on AD: (i) structural MRI, (ii) functional MRI (fMRI), and (iii) PET. As a fourth modality, diffusion tensor imaging

(DTI) bridges the gap between brain structure and function. These imaging methods are believed to reflect specific neurobiological changes of disease. Correlations between MRI-based volumetry and neuron numbers in specific brain regions suggest structural MRI as a potential marker of neuronal loss [6]. The blood oxygen level-dependent (BOLD) fMRI signal is primarily a measure of the input and processing of neuronal information within a region [7], opening an avenue to the assessment of intracortical functional connectivity. PET using ^{18}F-FDG is thought to represent neuronal glucose consumption as the main determinant of neuronal metabolism. Other PET ligands have been designed to bind AD characteristic fibrillar Aβ in amyloid plaques in vivo. DTI measures have been shown to reflect the integrity of white matter fiber tracts as the structural basis of functional connectivity. Three core criteria have been postulated for the use of a biological measure as diagnostic marker [4]: (i) applicability, i.e. the test should be non-invasive, widely available, and pose low patient burden, (ii) reliability, both within one center (test-retest reliability) and across centers, and (iii) validity in respect to underlying pathology and clinical outcomes.

The development of imaging biomarkers has entered a new phase with the conduction of large multicenter network studies, such as the North-American Alzheimers Disease Neuroimaging Initiative (ADNI) or the European ADNI (E-ADNI). These and other large-scale national and international multisite networks have allowed experimental markers to enter testing for practicability and multicenter stability along fast acquisition and analysis protocols in large and well-defined clinical cohorts. This renders it desirable to gain an overview about the present state of methodological developments and their potential applications in clinical research and clinical practice. An overview of important imaging markers is given in table 1, further details on these markers with specific emphasis on their potential use as diagnostic markers according to the core criteria of a useful biomarker is given in the following sections.

Structural MRI

In clinical trials, structural MRI markers of AD mainly serve to enrich, define and stratify an at-risk population for trials on potential disease-modifying or -preventive drugs. An academic expert stakeholder workshop convened by the EMEA in London (January 2010) seemed to reach an agreement on the introduction of neuroimaging, particularly structural MRI-derived biomarkers, alone or in combination with CSF candidates, for trial enrichment and enhanced diagnosis of AD at prodromal stages [8].

Structural MRI has already widely been explored in respect to the core criteria of applicability, reliability and validity for a useful biomarker. Structural MRI is non-invasive, can easily be repeated and is relatively widely available. The reliability of volumetric measures obtained from repeated MRI scans is generally high [9, 10]. Less is known about the variability of volumetric measures obtained from different MRI

Table 1. Imaging markers for AD

Marker	Diagnostic use	Comments
Visual rating of hippocampus volume	High correlation with hippocampus volume (R^2 ~0.9), diagnostic accuracy for AD vs. controls between 80 and 90%; prediction of AD in MCI not assessed	Useful for diagnostic evaluation at baseline; no use for follow-up
Manual volumetry of hippocampus	Diagnostic accuracy for AD vs. controls between 80 and 90%; prediction of AD in MCI with 70–80% accuracy [17]	Already employed in clinical trials as secondary endpoint, multicenter variability of longitudinal changes needs to be assessed
Automated volumetry of hippocampus	High correlation with manual volumetry ($R^2 > 0.8$). Group discrimination AD vs. controls 83%, MCI vs. controls 73% (post-hoc probability only) [21]	Evaluation in larger multicenter studies has just begun
Manual volumetry of entorhinal cortex	No additional benefit in identifying patients with manifest AD. Accuracy of prediction of AD in MCI increased by a few percent compared to hippocampus volumetry in monocenter studies [24, 294, 295], no additional use in first multicenter studies [28]	Diagnostic use for prognosis of AD in MCI seems to be superior to hippocampus, use as secondary endpoint in RCT appears not superior to hippocampus considering the laborious methodology
Automated measurement of whole brain volume	Diagnostic use only for the rate of change, not the baseline volume, but would require second scan after 1 year before diagnosis can be supported	Already secondary endpoint in clinical trials, but only limited heuristic value due to the global nature of the measurement
Voxel-based morphometry (VBM)	Characteristic pattern of brain atrophy in AD and MCI, but lacks an established statistical model to determine individual risk for a single subject. Combining VBM with regions of interest yields 78% sensitivity and 75% specificity in predicting MCI in healthy subjects, but only post-hoc probability	Combination of VBM with a region-of-interest approach allows diagnostic use, but abandons the advantage of mapping changes across the entire brain
Deformation-based morphometry	Statistical models for individual risk prediction have been proposed; predicts AD in MCI with 80–90% accuracy [53, 296], discriminates between MCI and controls with 90% accuracy [49]	Allows individual risk prediction based on information on atrophy pattern across the entire brain. Awaits confirmation in larger multicenter and longitudinal studies
Cortical thickness measurement	90% accuracy in the discrimination between AD patients and controls; reaches effect sizes similar to manual hippocampus volumetry when predicting AD in MCI [28]	Promising endpoint in clinical trials as it offers a direct assessment of effect sizes expressed in a meaningful metric

Table 1. Continued

Marker	Diagnostic use	Comments
DTI	High mean diffusivity values and low fractional anisotropy values within the left hippocampus predicted MCI in a mixed cohort of MCI and healthy elderly with a accuracy of 83 and 81%. Good effect sizes (Hedge's g from 0.5 to 1.14) in several brain areas (esp. posterior cingulate bundle, splenium, temporal white matter, parahippocampal, uncinate fascicle, and superior longitudinal fascicle) for the distinction of AD and healthy controls. Also good effect sizes for the discrimination of MCI and healthy controls	Promising for: – Early detection of AD – Quantification of cognitive reserve – Discrimination of age-related degenerative pattern and AD-related violation of structural connectivity – Multimodal approaches together with AD-specific imaging tools (amyloid PET). Application in larger multicenter and longitudinal studies has begun
fMRI with active paradigms	Episodic memory paradigms show consistent differences between AD patients and controls (reduced activation in AD). At-risk subjects showed increased and decreased activations in different studies, possibly relating to compensation mechanisms. Hyperactivation may predict faster cognitive decline	Promising for early detection of AD in predementia stages, but methodological questions of multisite stability need to be resolved. Connectivity-based analyses may add to the diagnostic use of fMRI acquisitions.
rfMRI impaired DMN functional connectivity	Allows separation between AD and control subjects with sensitivity ranging from 72 to 85% and specificity from 77 to 80%, shows potential to detect preclinical changes. Changes are present in preclinical stage, and predict conversion of MCI to full AD	Hippocampus and prefrontal cortex mostly affected, agrees with patterns of amyloid deposition, may predict the progression of MCI to AD
rfMRI decreased regional connectivity (ReHo), phase shift synchrony	Potential in improving statistical power of AD and MCI detection	Novel application of resting state functional connectivity (RSFC), may help in understanding of its underlying mechanisms
rfMRI graph theoretical analysis	Promising directions for better understanding of network characteristics that may lead to future biomarker	Loss of small-world properties in AD, with a notable reduction in the clustering coefficients
rfMRI coherence between anatomical (DTI) and functional connectivity	Coherence between two modalities decreases with age in healthy controls, but increases in MCI patients, thus showing trends in healthy aging reversed by MCI development	May support a novel mechanism of brain restructuring occurring in healthy aging

Table 1. Continued

Marker	Diagnostic use	Comments
FDG-PET	High accuracy of ≥90% for detection of AD, particularly when using automated statistical analysis. Reduced accuracy in older subjects (≥74 years) in carriers of APOE ε4 allele and in the differentiation of AD vs. DLB or progressive non-fluent aphasia. Guidelines of referring physicians state that FDG-PET can contribute to the clarification of unclear cases	Potentially very stable and valid multicenter marker, however, limited due to costs and availability
Amyloid PET	High sensitivity in detecting amyloid plaques and vascular amyloid in human brain in vivo. Prediction of cognitive decline in unimpaired subjects not yet resolved [210]	Potentially, interesting diagnostic marker pending further follow-up studies in PIB-positive healthy elderly subjects and MCI patients, not clear if useful as a surrogate endpoint. ^{18}F markers are presently undergoing phase II and III diagnostic trials
Cholinergic system PET (^{11}C-N-methyl-4-piperidyl-acetate (MP4A) and ^{11}C-N-methyl-4-piperidyl-propionate (MP4P))	For all currently available cholinesterase inhibitors at standard clinical dose the average reduction of cerebral AChE activity is in the range of 30–40%	Potentially interesting biomarker to determine the mode of action of a new compound

scanners. The variability of manual and automated volumetric measures across 12 different MRI scanners was below 5% in one study [11], the variability of manual volumetric measurement of the hippocampus was only 3.5%. This is in the range of accuracy of manual or automated segmentation protocols tested against simple or complex phantoms in a single-center approach [10], suggesting that multicenter variability, once minimal criteria for scanner quality are met by the participating centers, will not limit the application of structural MRI in clinical trials. The validity of MRI-based markers for the underlying disease pathology has only partly been established. Clinicopathological comparison studies have shown that hippocampus volume obtained ante-mortem accounted for at least 50% of variability in neuron numbers determined during autopsy [12]. The amount of variation explained by MRI-based hippocampus volumetry was above 90% when MRI scans had been obtained post-mortem [13]. The relationship of MRI-based atrophy measures to clinical outcome measures of AD such as cognitive decline have also been shown in several studies. The question is still open whether clinical and structural outcomes are closely related in multicenter studies as well.

Regional Volumetric Measures

Atrophy measures in MRI have mainly focused on the hippocampus due to its early involvement in AD and the relatively ease of identification and measurement of this structure in anatomical MRI scans. Visual rating of the hippocampus formation reaches an accuracy in cross-sectional studies that is comparable to the accuracy of more labor-intensive manual volumetry [14–16]. In contrast, the use of a visual rating scale to determine rates of atrophy over time only reached accuracy at chance level, suggesting that this approach may not be usefully employed as a secondary endpoint in clinical trials [16]. Using manual volumetric measures, significant atrophy of the hippocampal formation can be demonstrated by MRI even during prodromal stages of AD and predicts later conversion to AD with about 80% accuracy [17, 18]. Presently, a combined effort of the European Alzheimer's Disease Consortium (EADC) and the ADNI is underway to harmonize protocols for the manual tracing of the hippocampus with the goal to develop and validate a unified standard protocol (http://www.centroalzheimer.org/sito/ip_sops_e.php). Automated methods have been developed that account for more than 90% of variation in manual measurements and reduce the measurement time from 3 h down to 0.5 h [19, 20]. Automated protocols have been employed in diagnostic studies where diagnostic accuracy for separating AD patients, mild cognitive impairment (MCI) subjects or MCI converters from controls ranged between 73 and 88% [21, 22].

Volumetric studies beyond the hippocampus have focused on the entorhinal cortex. It is thought to be affected by the neurodegenerative process of AD at a particularly early stage. Cross-sectional studies have shown that entorhinal cortex volumetry is unlikely to provide any additional benefit over hippocampus volume in identifying patients with manifest AD [23–26]; however, at the MCI stage, it may improve prognostic efficiency by a few percent compared to hippocampal volumetry [24, 27], although one multicenter study suggested no additional benefit [28]. Another brain region finding increasing attention is the cholinergic basal forebrain. Here, first studies show a reduced volume of basal forebrain nuclei in AD and MCI compared to healthy controls [29–32]. The diagnostic accuracy of cholinergic forebrain atrophy, however, has been found to be inferior to the accuracy of hippocampus volumetry in the MCI stage of AD.

Automated Data-Driven Methods

Due to the laborious nature of manual volumetric methods, automated methods have been developed to determine changes in the brain structure of AD patients using hypothesis- and rater-independent approaches. The most widely published automated post-processing method to date is voxel-based morphometry (VBM). The method is based on a low-dimensional spatial transformation of brain scans into a common reference space to control for global differences in brain size and shape. After segmentation, differences in gray matter volumes remaining on a local scale after accounting for global differences are the parameters of interest that drive a voxel-based univariate statistic [33, 34].

In cross-sectional studies on patients with AD or MCI, VBM yields patterns of cortical and subcortical atrophy that are consistent with findings from manual volumetric studies and neuropathological examinations [35–41]. By automatically extracting gray matter volumes through predefined regions of interest from the gray matter maps in standard space, the VBM approach can be employed for diagnostic purposes [42]. This approach, however, abandons a major advantage of VBM, i.e. the mapping of changes across the entire brain. VBM still lacks validation in respect to the underlying neurobiological changes. The validity of VBM-based measurements of cortical atrophy in respect to potential clinical outcomes, such as memory performance, has begun to be established in a few studies so far [43, 44]. Combined with results from other neurodegenerative dementia disorders [45, 46], these data suggest that VBM reveals characteristic pattern of brain atrophy whose relation to specific cognitive impairments is consistent with a priori models of the anatomical representation of cognitive function in cortical and subcortical neuronal networks in the human brain.

While VBM transforms brain images into a standard space to compensate for global differences but preserves local differences in cortical gray matter distribution, deformation-based morphometry transforms the brain volumes at high resolution to a standard template to completely eliminate the anatomical differences between the brains. The anatomic information then resides in the deformation fields that are required to transform the patient's brain into a standard brain. These deformation fields offer a multivariate vector field of localization information from which regional volume effects can be extracted.

Several studies have impressively demonstrated the spread of atrophy through the brain in AD using high dimensional normalization [47, 48]. As in VBM, however, this approach faces the problem that it does not easily offer an individual measure of effect size that is needed to interpret the findings of a clinical trial.

Multivariate approaches offer a potential solution to this problem. Multivariate machine learning algorithms have been applied to deformation-based morphometric data to calculate an individual risk for the presence of AD in MCI subjects [49–53]. These algorithms reach up to 90% accuracy in identifying those MCI patients who convert into dementia after 2–3 years of follow-up [50, 52]. In one study, support vector machine-based classification correctly identified AD in 96% of pathologically confirmed cases and correctly separated between AD and frontotemporal dementia in 89% of pathologically confirmed cases [54].

Measurement of cortical thickness may offer an alternative solution to the problem of determining individually applicable measures of atrophy based on automated measurements. This approach yields values of cortical thickness at each location of the cortical mantle in millimeters [55, 56]. Group separation based on cortical thickness showed an accuracy of more than 90% in distinguishing between AD patients from healthy controls [57–59]. If such a marker could be established in multicenter trials and be validated against cognitive and neurobiological measures, it might prove very useful as risk marker for AD as it offers a direct assessment of individual changes expressed in a meaningful metric.

Summary of Volumetric MRI

Hippocampus volumetry currently is the best-established diagnostic biomarker candidate in structural neuroimaging for AD and prodromal MCI. The marker needs further validation in respect to the underlying neurobiological substrate and potential confounds such as vascular disease, inflammation, hydrocephalus and alcoholism, and in respect to clinical outcomes such as cognition but also to demographic and socioeconomic outcomes such as mortality and institutionalization. The diagnostic use of hippocampus volumetry will further increase with the broad availability of automated measurement approaches. Among the automated measurements, in the medium term, regionally differentiated methods will become available together with an appropriate statistical model, such as multivariate analysis of deformation fields, or techniques such as cortical thickness measurements that yield a meaningful metric.

Diffusion Tensor Imaging

DTI is increasingly recognized as a new MRI technique for quantifying the structural integrity of brain tissues [60, 61].

In AD, pathological white matter alterations have been reported, including decreased myelin density, loss of oligodendrocytes and microgila activation [62–64] with an increased vulnerability of later myelinating regions [65–67]. Additionally, frequent microvascular changes in AD [68] and wallerian degeneration of fiber connections due to cortical neuronal loss are likely to contribute to the white matter changes in AD [69]. White matter damage can be sensitively quantified by DTI that measures the random motion of water molecules [70, 71]. The mean diffusivity is a measure of randomized mean water diffusion. In neurodegenerative disease with a progressive loss of barriers that restrict water molecule motion (e.g. neuronal loss), an increase of mean diffusivity can be detected. Thus, mean diffusivity can be used to quantify the degree of degeneration of brain tissues. In structured tissues with myelin fibers this motion is restricted with a relative preference for the diffusion along the axial fiber direction (anisotropic diffusion). The degree of anisotropy is given as fractional anisotropy and can be calculated from the eigenvalues of the diffusion tensor (3-D ellipsoid model), which is used as a measure for white matter integrity [72–74]. With the advanced technological aid of DTI, MRI became capable of quantifying the structural integrity of biological tissues, especially CNS fiber tracts both in vivo and in vitro. A recent further development of DTI data processing is fiber tractography which allows the reconstruction of fiber pathways by following the orientation of maximum diffusion-tensor voxel by voxel [75].

Using region-of-interest-based DTI analyses in AD or MCI, several studies demonstrated brain structural disturbances predominately in regions that are known to be preferentially or early affected in AD (hippocampal, temporal, posterior cingulate bundles, parietal) [76–86]. Voxel-based analyses of DTI data consistently showed comparable results [87, 88], especially the white matter tracts within corpus callosum, the cingulum, and fornix, as well as the frontal and temporal white matter seem to be early affected in AD.

A recent meta-analysis of 41 published DTI studies in AD and MCI (including a total of 2,026 subjects, 617 of whom had AD, 494 with MCI, and 917 normal controls) found significant decreases of fractional anisotropy in AD compared to controls with moderate to strong effect sizes (Hedge's g from 0.5 to 1.14) in the posterior cingulate bundle, splenium, temporal white matter, parahippocampal, uncinate fascicle, and superior longitudinal fascicle [89].

Although there is evidence of a close relation between cortical neuronal loss in AD and subsequent degenerative changes of the connected white matter tracts [69], DTI is referring to distinct microstructural pathomechanisms and cannot be replaced by volume measures [90, 91].

In two independent longitudinal observation studies, baseline hippocampal diffusivity predicted conversion to dementia in MCI patients who were followed up by clinical assessment over a mean period of 1.5–3 years [92, 93]. In manifest clinical stages of AD, diagnostic accuracy of DTI-based measurement of fractional anisotropy and mean diffusivity changes using a multivariate network analysis was comparable to the diagnostic accuracy of classical MRI-based gray matter volume measurements [94]. Taking together these findings suggests that DTI measurements may be of particular diagnostic relevance in predementia stages of AD, however these findings require further replication.

Summary and Perspectives of DTI

DTI can already be regarded as powerful new MR technique to sensitively quantify alterations of white matter integrity. The utility and feasibility of DTI for early AD diagnosis and for the monitoring (or treatment) of the disease needs further evaluation, especially compared to other MR measures and in the context of larger multicenter approaches.

Fiber tractography, which allows to reconstruct and quantify fiber connections, could qualify as feasible new DTI data post-processing tool for automated analyses of functionally structural network components that are early affected in AD [95]. One further promising development is the combination of DTI with other imaging modalities, especially different PET modalities such as FDG-PET [96] or amyloid PET, and fMRI [97]. These studies are likely to improve our understanding of the interplay of AD specific pathology, synaptic (dys)function and underlying network integrity and thus detect necessary conditions of upcoming clinical AD. There is first evidence that DTI might contribute to quantification of individual cognitive reserve [98] and thus, the multimodal approach should improve our knowledge on individual resilience and vulnerability to AD pathology on the imaging level.

Functional MRI

Task Functional MRI in Alzheimer's Disease

fMRI has primarily been used as a cognitive neuroscience tool to probe the neural underpinnings of cognitive dysfunction in AD. To date, there are very limited data on fMRI test-retest reliability, cross-scanner platform reproducibility, and correlation with longitudinal clinical outcome available. Nevertheless, fMRI has several potential advantages as a biomarker in AD, particularly for 'proof-of-concept' studies in drug development and in detecting subtle brain dysfunction in very early stages of AD.

fMRI is a non-invasive imaging technique that does not require the injection of contrast agent, and thus can be repeated many times over the course of a longitudinal study or clinical drug trial. It has relatively high spatial and temporal resolution, and can be acquired in the same session as structural MRI. Perhaps most importantly, fMRI may provide useful information about the functional integrity of brain networks supporting memory and other cognitive domains, including the neural correlates of specific behavioral events, such as successful memory formation [99–102].

It is important to acknowledge that fMRI is an indirect measure of neuronal activity, inferred from measuring changes in BOLD MR signal [103, 104], but multiple studies suggest that fMRI activity closely approximates task performance. While FDG-PET is thought to be primarily a measure of synaptic activity, BOLD fMRI is considered to reflect the integrated synaptic activity of neurons via MRI signal changes due to changes in blood flow, blood volume, and the blood oxyhemoglobin/deoxyhemoglobin ratio [105]. Task fMRI studies typically compare the MR signal during one condition to another, either in blocks of stimuli (e.g. novel stimuli compared to familiar) or in event-related designs (e.g. stimuli that were correctly remembered compared to those that were forgotten).

Much of the early fMRI work in MCI and AD utilized episodic memory tasks, and was focused on the pattern of fMRI activation in hippocampus and related structures in the medial temporal lobe (MTL). In patients with clinically diagnosed AD, the results have been quite consistent, showing decreased MTL activity during the encoding of new information [106–114]. Neocortical fMRI changes have also been demonstrated in AD using fMRI, including decreased activation in temporal and prefrontal regions. A recent quantitative meta-analysis [115] of both fMRI and FDG-PET memory activation studies of AD identified several regions as consistently being more likely to show greater encoding-related activation in controls than in AD patients, including hippocampal formation, ventrolateral prefrontal cortex, precuneus, cingulate gyrus, and lingual gyrus. Interestingly, evidence of increased neural activity has been observed in AD patients performing memory tasks [110, 116].

The fMRI studies in subjects at risk for progression to AD, both MCI subjects and genetic at risk, have yielded much less consistent findings. Several studies have reported decreased MTL activation in MCI compared to controls [111, 117–119]. A number of studies in genetic at-risk subjects have also reported decreased MTL activity [120–125]. A slightly larger number of studies have reported evidence of increased MTL activity in at-risk subjects, particularly among very mild MCI subjects [126–131] and cognitively intact individuals with genetic risk for AD [132–138]. A common feature of the studies reporting evidence of increased fMRI activity is that the subjects were able to perform the fMRI tasks reasonably well. In particular, the event-related fMRI studies have reported that the hyperactivity was observed specifically during successful memory trials, suggesting that the increased activity may serve as a compensatory mechanism in the presence of early AD pathology [139, 140]. Cross-sectional studies suggest that the hyperactivity may be present only at early stages of MCI, followed by a loss of activation in late stages of MCI, similar to the pattern seen in AD patients [129]. Longitudinal clinical follow-up studies furthermore suggest that the presence of hyperactivity at baseline is a predictor of rapid cognitive decline [126, 132, 141], and loss of hippocampal function [142]. These longitudinal data suggest that hyperactivity may be a marker of impending neuronal failure. The mechanistic underpinnings of MTL hyperactivation remain unclear. This phenomenon may reflect cholinergic or other neurotransmitter upregulation [143], aberrant sprouting of cholinergic fibers [144], inefficiency in synaptic transmission [145], increased calcium influx or excitotoxicity [146, 147]. Further research to determine the specificity of hyperactivation with respect to particular brain regions and behavioral conditions, and the relationship to baseline perfusion and metabolism will be valuable to better characterize this phenomenon.

Converging data suggest that memory function is subserved by a network of brain regions, which includes not only the MTL system, but also a set of cortical regions, including the precuneus, posterior cingulate, lateral parietal, lateral temporal and medial prefrontal regions, collectively known as the 'default network' which typically deactivate during memory encoding and other cognitively demanding tasks focused on processing of external stimuli [148, 149]. Recent studies have also suggested that the default network demonstrates markedly abnormal responses during memory tasks in clinical AD patients and in subjects at risk for AD [129, 150–154]. Interestingly, the same default network regions that typically demonstrate beneficial deactivations in healthy young subjects, especially the posterior cingulate/precuneus, tend to manifest a paradoxical increase in fMRI activity above baseline in AD patients. Functional connectivity studies, both at rest and during task, have consistently demonstrated evidence of disrupted connectivity within the default network in AD [155], and more recently in MCI patients destined to progress to AD dementia [156].

Interestingly, the default network regions demonstrating aberrant task-related fMRI activity and dysconnectivity in MCI and AD also overlap the anatomy of regions with highest amyloid burden in AD patients [157–159]. Several recent studies in cognitively normal older individuals with evidence of amyloid deposition on PET imaging have demonstrated evidence of disrupted default network activity during memory tasks and at rest [140, 160, 161] suggesting this combination of imaging biomarkers may be particularly useful to detect very early brain dysfunction in preclinical AD.

fMRI may hold the greatest potential in the evaluation of novel treatments for AD. Several studies in healthy young and older subjects suggest that fMRI can detect acute pharmacological effects on memory networks [162–164]. To date, only a few small fMRI studies have demonstrated enhanced brain activation after acute or prolonged treatment with cholinesterase inhibitors in MCI and AD, although these studies were not conducted as typical double-blind, placebo-controlled trials [165–170]. More recently, fMRI has been used to evaluate the effects of behavioral interventions, such as mnemonic training [171]. fMRI has also been incorporated into a small number of investigator-initiated add-on studies to ongoing phase II and phase III trials, which should provide some valuable information regarding the potential utility of these techniques in clinical trials. There are, however, several challenges in performing longitudinal fMRI studies in patients with neurodegenerative dementias. It is likely that task fMRI will remain quite problematic in examining patients with more severe cognitive impairment, as these techniques are very sensitive to head motion. If the patients are not able to adequately perform the cognitive task, one of the major advantages of task fMRI activation studies is lost. Resting state fMRI may, however, be performed with more severely impaired patients.

In terms of using fMRI in longitudinal or pharmacological studies, it is critical to complete further validation experiments. BOLD fMRI response is known to be variable across subjects. The short-term reproducibility of BOLD signal changes within young healthy individuals during memory encoding tasks and resting fMRI is reported to be reasonable [172–174]. However, very few studies examining the reproducibility of fMRI activation in older and cognitively impaired subjects have been published to date [175] [176]. Longitudinal functional imaging studies are needed to track the evolution of alterations in the fMRI activation pattern over the course of the cognitive continuum from healthy aging to clinical dementias such as AD. It is also important to evaluate the contribution of structural atrophy to changes observed with functional imaging techniques in neurodegenerative diseases. Finally, longitudinal multimodality studies, including structural MRI, fMRI and FDG-PET and PET amyloid imaging techniques, are needed to understand the relationship between these markers and the relative value of these techniques in tracking change along the clinical continuum of AD [177].

In summary, much of the task fMRI work in early AD to date has been primarily in the realm of cognitive neuroscience and in elucidating the mechanistic underpinnings of memory impairment. Additional work to validate fMRI as a potential biomarker for use in large-scale trials is critically needed. Recent work using fMRI, which does not require special equipment or ability to perform a task, suggests that these techniques may be particularly amenable to use in clinical trials. It is likely that task fMRI and fMRI may have the greatest utility in early proof-of-concept studies, to detect a signal of efficacy over a relatively short time frame, particularly in individuals at very early stages of AD, which may predict long-term disease modification.

Resting State Functional Connectivity

The Method

In fMRI time series collected while subjects are not performing any explicit task, the spontaneous fluctuations in images sensitive to blood oxygenation (BOLD) are strongly correlated between distant regions of several functional systems. First observed in motor regions, these findings of resting state functional connectivity (RSFC) were extended later to show that analysis of resting state data allow detection of several functional networks, both ones that were previously known, such as sensory, language and attention systems, and new ones such as the default mode network, and its anticorrelated analogue, the task positive network. The true psychological origin of temporal correlation is still not well understood, but multiple evidence has excluded physiological noise as a sole origin of those correlations and established connection between resting state connectivity and other indicators of brain connectivity obtained using electrophysiology, structural imaging and network analysis of fMRI and PET activation studies. The typical resting state connectivity study is simple to perform, requiring only one 5- to 10-min long resting imaging run while the subject is instructed to rest (with eyes closed or fixated on the crosshair). Therefore, these studies give access to patients in advanced stages of AD that cannot comply with the requirements of activation fMRI studies.

The information derived from experiments in resting state connectivity allows detecting the presence of correlations between various brain regions, and quantifying the strength of those correlations. Main analytical methods can be divided into model-driven, mainly seed-based, and data-driven. In the model approach, spatial analysis is performed via seed region analysis where the time course of a voxel or brain region is selected and the time course of each brain voxel is correlated with that seed time course. Alternatively, data-driven analysis can be performed, usually in the form of independent component analysis when the whole dataset is analyzed to extract brain components characterized by a similar time course [178–180]. Both model- and data-driven methods allow for later quantification of connection strength for group comparison.

The network most commonly investigated in resting studies that happens to be also implicated in AD is the default mode network (DMN). It is a system of regions that have been identified as negatively activated in many cognitive tasks. It contains regions of posterior cingulate cortex, medial prefrontal cortex, medial temporal lobe, parts of the precuneus, lateral and inferior parietal cortex and hippocampus, typical predilection sites of AD pathology [181, 182].

Connectivity Decreases in Alzheimer's Disease

RSFC was found to be decreased in MCI patients compared to healthy controls, and still lower in AD patients [183]. The deactivations in DMN were found to be decreased in AD and functional connectivity between hippocampus, posterior cingulate cortex and other regions of DMN was found to be affected [184–188]. These findings were significant enough to suggest resting fMRI of DMN connectivity as a biomarker for separation between clinical AD and controls with sensitivity ranging between 72 and 85% and specificity between 77 and 80%. Not all resting connectivity measures are lower in AD. A global network-based analysis by Sorg et al. [189] has shown that out of eight resting state networks identified in a large resting state study, only two, DMN and executive networks, exhibited decreased connectivity in AD. The lowered functional connectivity in AD is accompanied by less pronounced but still significant decrease in MCI [190, 191].

The changes in DMN connectivity in AD are of great interest because they are present already in predementia stages. Petrella et al. [192] have followed a group of MCI patients over 2–3 years and have shown that DMN connectivity was lower in patients who subsequently progressed to the diagnosis of AD compared to those who did not decline. This further supports hopes that DMN connectivity may be used to develop marker for preclinical changes leading to AD.

Relation to Amyloid Depositions and Genetics of Alzheimer's Disease

The resting connectivity measures have been compared with estimates of amyloid deposits that are thought to underlie the AD pathology [181, 193]. Use of PIB (Pittsburg Compound B), a marker of amyloid plaque accumulation in the brain, has shown that resting connectivity of DMN was lower in patients with heavy amyloid burden. This may even suggest an explanation why DMN shows the strongest effect of AD on connectivity as its regions have been found to contain regions with the highest concentration of amyloid depositions. Furthermore, clinically normal subjects with a higher level of amyloid deposits have been shown to have lower connectivity in DMN, suggesting that resting connectivity may be able to capture preclinical stages of AD development. The analysis of genetic background of AD and resting state connectivity by Filippini et al. [194] has shown that young carriers of APOE ε4 allele risk factor showed a surprisingly higher connectivity in DMN. While seemingly contradicting earlier findings of aging populations, this suggests that the function of DMN regions subject to the disease process in AD are modulated by the presence of the *APOE* 4 allele even in young adults [182].

Other Resting State Findings in Alzheimer's Disease

There are several more exploratory methods of analysis of resting state data, and some of them have been successfully applied to AD and MCI. Among those, only few have been applied to the study of diagnostic use in AD as a biomarker, which are briefly described below.

Graph theoretical analysis has been applied to analyze properties of networks revealed when multiple (up to a 100) regions are anatomically defined and its connectivity is estimated using RSFC [195]. Its application to AD [187] has shown a loss of small-world properties in AD, with a notable reduction in the clustering coefficient – a sign of reduced local connectivity. The statistical power of this techniques points to a potential for using it as a biomarker.

Another approach investigating brain connectivity has been proposed by Skudlarski et al. [196] in an analysis of the spatial coherence between functional and anatomical connectivity and its aging trajectory. DTI fiber tracking is combined with RSFC to create connectivity matrices for both modalities that can be directly compared. The analysis of similarity (spatial coherence) between those two matrices shows that age-related patterns differ between healthy aging and MCI patients [197]. While the coherence between those measures decreases with age in healthy aging it is increasing in patients. This suggests that in the healthy aging brain, functional connectivity is being reorganized to cope with the deterioration of white matter, and thus resting and structural connectivity are decoupled. It may be the lack of this reorganization and the decoupling between connectivity measures following it that marks the development of MCI and AD. Those results would support the hypothesis of adaptive neurocognitive scaffolding introduced by Park and Reuter-Lorenz [198].

Limitations

The main limitation of RSFC analysis is its weak physiological underpinning. While results of resting state analysis are robust and in agreement with other modalities such as electrophysiology, MEG, and anatomical connectivity measurements, it is not clear what physiological differences on neuronal level are causing changes in resting state connectivity. Multimodal studies combining RSFC with other measures such as anatomical data from DTI, electrophysiological data from EEG and direct recording of evoked potentials and electrical stimulation will hopefully allow more complete description of brain connectivity and will be able to address which components of connectivity are affected by AD and by what mechanisms, thus allowing to advance from detection of deterioration, that is clearly in our reach now, to slowing or preventing it.

Positron Emission Tomography

PET utilizes biologically active molecules in micro- or nanomolar concentrations that have been labeled with short-lived positron emitting isotopes. The physical characteristics of the isotopes and the molecular specificity of labeled molecules, combined with the high detection efficacy of modern PET scanners, provide sensitivity for detection of specific molecules in brain tissue that is several orders of magnitude higher than with other imaging techniques. Whereas the very short half-live of carbon-11 (^{11}C, half-life 20 min) limits its use to research applications in fully equipped PET centers with a cyclotron and radiopharmaceutical laboratory, fluorine-18-labeled tracers (^{18}F, half-life 110 min) can be produced in specialized centers and distributed regionally to hospitals running a PET scanner only (table 1). Clinical use of PET is now well established in clinical oncology and it has the potential for becoming more widely available, also for application in AD [199, 200].

Amyloid Imaging

The deposition of amyloid-β (Aβ) is an early event in the pathogenesis of AD and is central in the amyloid cascade hypothesis. The possibility to assess amyloid imaging in vivo by PET has therefore attracted much interest. The first tracer to be used to label fibrillary Aβ selectively with high affinity in vivo was ^{11}C-labelled Pittsburgh Compound B (^{11}C-PIB) [201]. In addition, unspecific binding is observed mainly in white matter. Many studies have demonstrated that this tracer has a very high sensitivity of 90% or higher to detect fibrillary amyloid plaques in patients with AD [202–205]. Most likely a significant proportion of the up to 10% PIB-negative AD patients in clinical series will be due to clinical misdiagnoses. Only under exceptional circumstances has PIB negativity been confirmed in AD [206].

Most normal control subjects exhibit very low cortical binding of PIB, but a proportion of normal elderly controls are showing higher cortical PIB binding, typically resulting in a bimodal distribution of PIB uptake in samples of control subjects. The frequency of increased cortical PIB binding in controls increases rapidly from 20% or less below age 70 to 30–40% at age 80, largely reflecting similar findings in previous autopsy studies [207]. The clinical implications of amyloid deposition in elderly controls are not yet clear. Some of them also have indicators of beginning neurodegeneration [208, 209] or will develop cognitive deficits [210], but cerebral function in others may be resistant to amyloid deposition. The APOE 4 allele is a genetic risk factor for increased PIB uptake in cognitively healthy subjects [211].

Findings in patients with MCI are heterogeneous. In most studies, approximately two thirds of patients showed increased binding like AD patients, while the rest were within normal limits. Published results from follow-up studies indicate that patients with increased binding have a high risk for progressing to AD with manifest dementia [212, 213], while MCI patients with negative PIB scans very rarely develop dementia [214]. Patients with amnestic MCI show more PIB binding than non-amnestic MCI [215]. Amyloid deposition is high in posterior association areas, where it correlates with a decline of glucose metabolism, but it is also high in frontal association cortex where that correlation is absent [216]. In contrast to structural imaging and FDG-PET, there is no acceleration of changes with disease progression [217], and some studies even indicate no further increase of PIB binding once patients have become demented [199].

Amyloid imaging is expected to provide excellent differentiation of AD from frontotemporal dementia, which is not associated with amyloid deposition and increased ^{11}C-PIB binding [218]. Dementia with Lewy bodies (DLB) often also show fibrillary amyloid deposition in pathological studies, and correspondingly positive PIB scans were reported in most patients [219, 220]. Moderately increased PIB binding, predominantly in occipital regions and on average less than in AD, was also observed in non-demented patients with cerebral amyloid angiopathy [221].

There are currently three ^{18}F-labelled tracers in clinical trials, which have been developed as proprietary tracers for commercial distribution. These are flutemetamol (GE-067, 3′-fluoro-PIB), florbetaben (BAY-94-9172, AV-1), and florbetapir (AV-45). Flutemetamol, florbetaben and florbetapir appear to have largely similar imaging properties, although the optimum scanning time after intravenous injection varies. Results from clinical trials indicate that they likely will provide high diagnostic power for discrimination between AD patients and controls [222–224]. Cortical binding of ^{11}C-PIB and flutemetamol in cortical regions is closely correlated [225]. A phase 3 study of florbetapir in 29 patients with terminal disease who underwent PET on average 99 days before death demonstrated 96% correspondence of patient classification between PET and autopsy, and a good correlation between regional PET amyloid load and immunohistochemistry [226].

Glucose Metabolism

Cerebral glucose metabolism is being measured by the most widely available PET tracer ^{18}F-2-fluoro-D-deoxyglucose (FDG). There is close coupling of glucose metabolism with neuronal function [227]. Glucose is the main substrate for energy production which is required to maintain the neuronal ion gradients for neuronal activity and coupling to synaptic activity is also mediated by the neuron-astrocyte glutamate shuttle [228].

Over more than 20 years multiple studies have demonstrated that glucose metabolism and blood flow is impaired in temporal-parietal association cortices including the precuneus region and the posterior cingulate cortex very early during disease progression [for review, see 229]. In patients with manifest dementia, the angular gyrus usually being most severely affected, and the annual decrease of metabolism in temporoparietal association cortices is 5–6% [230]. Automated objective image evaluation software [231, 232] can provide objective tools for detection of these metabolic impairments, providing a diagnostic sensitivity of 80–95% at 65 to 90% specificity for diagnosis of AD [233, 234].

Microglial Activation

Microglia are the resident immune cells of the brain. In response to brain damage, microglia undergo changes in their morphology, migrate toward the lesion site, proliferate, produce cytokines and reactive oxygen species. This is associated with expression of the peripheral benzodiazepine receptor, which is known to be located at the mitochondrial translocator protein [235]. Activated microglia are present at sites of aggregated Aβ deposition in the brains of AD subjects [236], however the potential influence of microglia activation on disease progression remains still unresolved [237].

The first tracer that became available for imaging of microglial activation in humans has been ^{11}C-PK11195 [238, 239]. The in vivo PET findings using ^{11}C-PK11195 in AD are not particularly clear. An early study using racemic ^{11}C-PK11195 was negative [240], probably due to the relatively high level of non-specific binding resulting in unfavorable signal strength, while a recent study using the R-isomer according to current standards [241] found moderately increased binding. Thus, there is a need for development of better tracers, ideally labeled with ^{18}F for clinical use, and a large number of new tracers have been tested in experimental animals [242].

Cholinergic Neurotransmission

Pathological studies suggest a severe loss of cholinergic fibers and their characteristic enzymes and receptors in AD and DLB [243]. While there are no suitable tracers for acetylcholine transferase, tracers have been developed for other cholinergic markers. Labeled analogues of acetylcholine, which are also substrates for AChE, can be used to measure and image acetylcholine esterase (AChE) activity in vivo. These are ^{11}C-N-methyl-4-piperidyl-acetate (MP4A, also known as AMP) [244], which is 94% specific for AChE in human brain, and ^{11}C-N-methyl-4-piperidyl-propionate (MP4P, or PMP) [245]. A significant decrease of cortical AChE activity has been observed in MCI and AD [for review, see 246], probably reflecting the loss of AChE that is associated with cholinergic axons [247]. It is most severe in temporal neocortex, where it is correlated with memory deficits, while in other brain areas it is mostly related to deficits in attention [248].

Nicotinic receptors have attracted intense interest but available tracers still suffer from methodological limitations. ^{11}C-nicotine has a high level of unspecific binding, although reduced binding in AD can be detected [249, 250]. The a4b2 receptor subtype has been imaged using ^{18}F-A85380 [251, 252] and ^{131}I-A85380 [253], and reduction of binding has been observed in MCI and AD [254] as well as in Parkinson's disease with cognitive impairment [255]. However, binding kinetics are too slow for reliable quantitation and clinical use, which motivates ongoing research for better ligands [256].

Dopamine

The most widely used tracer to examine dopamine synthesis and vesicular storage is ^{18}F-fluorodopa [257]. A deficit of dopamine synthesis similar to Parkinson disease has been found in DLB, even at a stage when parkinsonism may not yet be prominent [258], while it is normal in patients with AD and therefore provides an important diagnostic marker. In contrast to the cholinergic impairment which is severe in DLB but only mild in Parkinson disease without dementia, the dopaminergic deficit does not appear to be related to dementia [259].

Summary of PET

PET provides tools for imaging of specific pathophysiological processes that lead to ultimate manifestation of dementia in AD. ^{18}F-labeled tracers can be used clinically and are of particular interest for early and differential diagnosis. FDG for assessment of regional synaptic function is already widely available, ^{18}F-labeled ligands for amyloid and dopamine transporters are under development by commercial companies. Tracers for the cholinergic system are being used in research studies to investigate potential treatment opportunities to improve attention and memory in dementia, and tracers for imaging of microglial activation will provide insight into the interaction between neuroinflammation and neurodegeneration.

Multicenter Imaging Studies

Standardization

Identification of atrophy on MR was proposed many years ago as an adjunct diagnostic tool in AD. Initial methods proposed were both qualitative [260, 261] and quantitative [262], and focused on detection of medial temporal atrophy (especially hippocampal) as an early marker of AD. Considerable technical progress has been made in the interim including not only dramatic improvements in MRI technology but also development of fully automated quantitative imaging analysis methods [21, 263–271] and multivoxel analysis methods that incorporate information not just from a single structure like the hippocampus, but from multiple brain areas [54, 272–275]. However, despite considerable technical progress, quantitative measures of MRI atrophy have not gained widespread clinical use. A major reason, perhaps the major reason, is the absence of widely accepted standardized methods [276]. Standardization is an essential component of any diagnostic method whether the context of use is a therapeutic trial or for clinical diagnosis.

Implicit in the notion of widespread acceptance and standardization is a multisite context of use. Standardization of an MRI acquisition requires that it be consistent across multiple sites and over time. That is, similar image qualities (contrast-to-noise, spatial resolution, resistance to artifact, reliability, speed, etc.) must be achieved across sites and vendor platforms over time. There are many components to standardization, but the initiating step is standardization of the acquisition protocol. While MRI technologists can manually enter parameters from a written document into a scanner and save the protocol, a far more efficient and error free method is electronic distribution of the protocol. This is how MRI vendors distribute protocols and software updates and is the preferred method for multisite studies.

Content of a Standardized Acquisition

Context of use obviously determines the content of any protocol. The requirements for a multisite research protocol differ somewhat from those of a routine clinical protocol. For example, while an experienced radiologist often can 'read-through' minor artifacts, the automated software programs used for quantitative analysis typically cannot. In fact, as a general rule, the more fully automated the analysis algorithm the less tolerant it is to image imperfections. Nonetheless, an important feature of a multisite research protocol is that it should contain sequences that are suitable for general diagnostic purposes. That is, a research protocol must enable identification of clinically significant abnormalities in order to serve the best interest of a subject who might be found to have an unknown but medically significant abnormality. Many IRBs now require that research examinations be interpreted in a customary clinical manner, and many in the field believe that this is an ethical imperative.

Imaging time must be considered in choosing specific sequences and sequence parameters for a multisite protocol. Many if not most elderly subjects find it difficult to tolerate examinations that greatly exceed about 45 min. Also, clinical MRI systems are typically scheduled in fixed 'slot times'. Scans are not scheduled and billed by the minute but in increments equal to the length of a typical examination. This is necessary for efficiency and for appropriate patient scheduling. A site survey for ADNI found that the typical 'slot' time was 45 min [277]. Hence multisite protocols that exceed 45 min may be problematic.

For research studies the protocol must contain the sequences needed to accomplish the aims of the study. A comprehensive protocol for an AD study should include several 'core' sequences including a 3D T_1-weighted volumetric scan for quantitative morphometric analyses. A scan with roughly 1-mm cubed resolution is possible with modern scanners. The 'core' protocol should also include a sequence for pathology detection and vascular disease evaluation. The typical recommendation would be FLAIR. FLAIR has the advantage over traditional T_2/spin density fast-spin echo imaging of providing a T_2-weighted scan but with the CSF signal nulled. A fast-spin echo T_2 sequence might be added in addition to FLAIR if a bright CSF acquisition is called for by the image analysis methods. A sequence that is now frequently used in AD studies is for detection of cerebral microbleeds. This can be accomplished either with a standard T_2* gradient echo sequence, or with a susceptibility weighted acquisition [278]. The latter is more sensitive for detection of microbleeds, but often requires a purchasable software key. In addition, standardization across MRI vendors is problematic for susceptibility weighted sequences. A standard T_2* gradient echo is more easily standardized for multisite studies.

Other imaging sequences of interest are resting state, arterial spin labeling (ASL) perfusion, and diffusion. Resting state requires only a task-free echo planar-type acquisition (EPI-BOLD) which can be executed on virtually any modern scanner, and hence can be readily standardized across MR vendors. DTI is much more difficult to standardize across vendor platforms, many would say impossible without using non-product imaging sequences. At this time, ASL perfusion imaging cannot be standardized across vendors without using non-product sequences (one of the major MRI vendors does not offer a product ASL at this time although that will change soon).

Most scanners can operate in either clinical mode or research mode. Research mode typically bypasses some of the more stringent limitations on scanner performance. A scanner operating in clinical mode can only execute imaging sequences and parameter ranges that are provided by the MRI vendor as a product that has cleared regulatory approval. Scanners operating in research mode offer more technical flexibility and higher performance specifications, but this requires that an active research agreement be in place between the site and the MRI vendor. Performing a large multisite study with sequences that must run in research mode is risky because the study is dependent on legal agreements between each site and the MRI vendor be active throughout the duration of the study. This is one of the reasons that the ADNI-2 study has specifically limited itself to imaging sequences that are approved product, run in clinical mode, and that do not require a research agreement [279]. A related consideration is the availability of specific software licenses, or keys, at the individual enrollment sites. More advanced imaging sequences like ASL, DTI, and even resting state EPI-BOLD, require that the site have purchased a software license key. These keys can run into the tens of thousands of dollars apiece. Therefore, consideration must be given to the availability of software keys across the fleet of scanners that must be supported in a study. It would make no sense, for example, to include ASL in a protocol if 80% of sites did not have the appropriate software key.

Other considerations for selecting the content of a multisite protocol include the fact that a wide variety of scanners must be supported. This may mean that those vendor systems with low market penetration may not be supportable. The study also may not be able to support older systems, even within a major vendor product line. The 'lowest common denominator' problem will affect most studies to some extent. In order to standardize the protocol across sites it is often not possible to take full advantage of the hardware/software capabilities of the most advanced systems, because the performance specifications of the most advanced systems cannot be matched by the older systems in the study.

Post-Acquisition Image Corrections

An optional strategy that may enhance standardization across sites and platforms is post-acquisition correction of certain image artifacts. These include corrections in image geometry for gradient non-linearity – i.e. 3D gradwarp [280, 281], corrections for intensity non-uniformity due to non-uniform receiver coil sensitivity [282], and correction of image intensity non-uniformity due to other causes such as wave effects at 3T. These corrections are system-specific. For example, the implementation of 3D gradwarp (or equivalent) is specific for each gradient configuration. Similarly, while correction for intensity inhomogeneity due to non-uniform sensitivity of multiarray receiver coils is now standard on modern systems, some older systems that might be included in multisite studies do not provide this feature. The ADNI study demonstrated that longitudinal consistency is improved with correction of scaling [283], gradient non-linearity [284], and with intensity non-uniformity correction [285, 286].

Quality Control

An essential feature of any multisite study is quality control. This is best accomplished by a central core MRI laboratory with the appropriate expertise and can be accomplished by visual, automated, or both image quality checks. Each examination done in the study should be examined by qualified personal in the central laboratory for protocol compliance and scanning errors (e.g. anatomic coverage). Each upgrade to a scanner runs the risk of corrupting the study protocol that was stored on the scanner. This is true for even minor software patches, so quality control should be done in near real time throughout out the duration of any multisite study.

An additional element of quality control that might be considered is monitoring individual scanners with a standardized phantom. ADNI demonstrated the value of scanner monitoring with a specially designed phantom; 20% of all ADNI-1 scans would have been affected by errors of various types had each scanner not been monitored [284]. Monitoring scanner performance with a phantom is a relatively inexpensive way to ensure that scanner errors are discovered and appropriate corrective action is taken before many errant scans have entered the study database.

In summary, with attention to the methodological details outlined above, highly reproducible quantitative MRI measures have been achieved in multisite studies [287, 288]. These studies demonstrate that quantitative MRI can in theory serve as a broadly distributed diagnostic measure for AD, not only in therapeutic trials but also for clinical purposes. However, first consensus must be achieved on standardized methods of acquisition, quality control and analysis. The lack of internationally agree upon standards poses the greatest barrier to implementation of quantitative MRI as an AD biomarker [289].

Which Imaging Marker to Use – A Clinical Perspective

The majority of clinical criteria for the diagnosis of AD use structural imaging to exclude other causes of dementia. Positive imaging markers of disease are still a matter of highly specialized clinical centers or experimental studies. The present state of imaging technology, as outlined in this chapter, offers the perspective for a broader application of positive imaging markers of disease in the near future.

Since structural imaging is generally part of clinical protocols, for diagnostic purposes the use of semiquantitative rating of MTL volumes will enhance diagnostic accuracy in early stages of AD at relatively low additional costs. The use of this technique for follow-up cannot be advised to date. More elaborated analyses, such as manual or automated volumetry or multivariate analysis of deformation-based maps, yield promising results in clinical trials, but there is not yet enough evidence to date to advise their use for clinical routine application.

The gold standard of clinical diagnosis to date is FDG-PET. However, the cost-effectiveness of FDG-PET has been questioned based on the limited effectiveness of treatments and the high costs of PET [290]. With the availability of disease-modifying treatments, the use of PET may become cost-saving when applied to allocate treatment to at-risk subjects.

For the near future we expect three major areas of application of ^{18}F-labeled markers for amyloid load outside of research. First, there will be a fraction of highly affluent middle-aged people who will request amyloid PET to confirm the absence of amyloid in their brains. Physicians who offer these diagnostics will have to decide how they will deal with positive findings given the absence of available treatment options for asymptomatic amyloid load and even the absence of firm evidence of an increased risk of dementia for middle-aged healthy subjects who show amyloid load in their brains. Second, a clinical indication will be the study of patients with a difficult differential diagnosis. In these patients, ^{18}F-amyloid PET will contribute to decide on individual treatment options. We still do not have much evidence, however, on the clinical relevance of positive amyloid load in patients presenting with cognitive impairment patterns that are atypical for AD. The third application will be patients with MCI. It is likely that well-informed MCI patients who have positive amyloid load will actively request treatment with antidementive drugs even outside of regulatory approval. This will pose a challenge both to physicians who will have to advise their individual patient and to political decision-makers and health insurances that will have to cover the expenses of both the PET study and subsequent treatment. Due to cost restrictions, we expect that access of patients with a classical clinical picture of AD-type dementia to amyloid PET will be very limited.

One question that needs to be resolved is the difference in diagnostic power between amyloid PET and measurement of tau and Aβ markers in CSF. Although being less invasive, the cost efficacy of amyloid PET relative to lumbar puncture needs to be established, where both the final costs of ^{18}F-labeled amyloid PET tracers and the diagnostic accuracy of amyloid PET compared to CSF tau and Aβ are still unresolved.

fMRI has no clinical application to date; this technique has not yet been applied in multicenter clinical trials. Resting state fMRI maps the same brain areas that are found affected in FDG-PET examinations. Resting state fMRI is easily available through a resting state sequence of 6–10 min duration. Although first studies suggest stability of resting state fMRI derived connectivity pattern [291], this still awaits confirmation in larger samples and under different disease conditions. We are just beginning to explore the potential of this technology.

References

1 McKhann G, Drachman D, Folstein M, Katzman R, Price D, Stadlan EM: Clinical diagnosis of Alzheimer's disease: report of the NINCDS-ADRDA Work Group under the auspices of the Department of Health and Human Services Task Force on Alzheimer's disease. Neurology 1984;34:939–944.

2 Dubois B, Feldman HH, Jacova C, Dekosky ST, Barberger-Gateau P, Cummings J, et al: Research criteria for the diagnosis of Alzheimer's disease: revising the NINCDS-ADRDA criteria. Lancet Neurol 2007;6:734–746.

3 Dubois B, Picard G, Sarazin M: Early detection of Alzheimer's disease: new diagnostic criteria. Dialogues Clin Neurosci 2009;11:135–139.

4 Hampel H, Frank R, Broich K, Teipel SJ, Katz RG, Hardy J, et al: Biomarkers for Alzheimer's disease: academic, industry and regulatory perspectives. Nat Rev Drug Discov 2010;9:560–574.

5 Hampel H, Wilcock G, Andrieu S, Aisen P, Blennow K, Broich K, et al: Biomarkers for Alzheimer's disease therapeutic trials. Prog Neurobiol 2011;95: 579–593.

6 Nagy Z, Jobst KA, Esiri MM, Morris JH, King EM-F, MacDonald B, et al: Hippocampal pathology reflects memory deficit and brain imaging measurements in Alzheimer's disease: clinicopathological correlations using three sets of pathologic diagnostic criteria. Dementia 1996;7:76–81.

7 Logothetis NK: The neural basis of the blood-oxygen-level-dependent functional magnetic resonance imaging signal. Philos Trans R Soc Lond B Biol Sci 2002;357:1003–1037.

8 Hampel H, Broich K: Enrichment of MCI and early Alzheimer's disease treatment trials using neurochemical and imaging candidate biomarkers. J Nutr Health Aging 2009;13:373–375.

9 Giedd JN, Kozuch P, Kaysen D, Vaituzis AC, Hamburger SD, Bartko JJ, et al: Reliability of cerebral measures in repeated examinations with magnetic resonance imaging. Psychiatry Res 1995;61: 113–119.

10 Byrum CE, MacFall JR, Charles HC, Chitilla VR, Boyko OB, Upchurch L, et al: Accuracy and reproducibility of brain and tissue volumes using a magnetic resonance segmentation method. Psychiatry Res 1996;67:215–234.

11 Ewers M, Teipel SJ, Dietrich O, Schonberg SO, Jessen F, Heun R, et al: Multicenter assessment of reliability of cranial MRI. Neurobiol Aging 2006;27: 1051–1059.

12 Zarow C, Vinters HV, Ellis WG, Weiner MW, Mungas D, White L, et al: Correlates of hippocampal neuron number in Alzheimer's disease and ischemic vascular dementia. Ann Neurol 2005;57:896–903.

13 Bobinski M, de Leon MJ, Wegiel J, Desanti S, Convit A, Saint Louis LA, et al: The histological validation of post-mortem magnetic resonance imaging-determined hippocampal volume in Alzheimer's disease. Neuroscience 2000;95:721–725.

14 Scheltens P, Leys D, Barkhof F, Huglo D, Weinstein HC, Vermersch P, et al: Atrophy of medial temporal lobes on MRI in 'probable' Alzheimer's disease and normal ageing: diagnostic value and neuropsychological correlates. J Neurol Neurosurg Psychiatry 1992;55:967–972.

15 Wahlund LO, Julin P, Johansson SE, Scheltens P: Visual rating and volumetry of the medial temporal lobe on magnetic resonance imaging in dementia: a comparative study. J Neurol Neurosurg Psychiatry 2000;69:630–635.

16 Ridha BH, Barnes J, van de Pol LA, Schott JM, Boyes RG, Siddique MM, et al: Application of automated medial temporal lobe atrophy scale to Alzheimer disease. Arch Neurol 2007;64:849–854.

17 Wang PN, Lirng JF, Lin KN, Chang FC, Liu HC: Prediction of Alzheimer's disease in mild cognitive impairment: a prospective study in Taiwan. Neurobiol Aging 2006;27:1797–1806.

18 Jack CR Jr, Petersen RC, Xu YC, O'Brien PC, Smith GE, Ivnik RJ, et al: Prediction of AD with MRI-based hippocampal volume in mild cognitive impairment. Neurology 1999;52:1397–1403.

19 Csernansky JG, Wang L, Swank J, Miller JP, Gado M, McKeel D, et al: Preclinical detection of Alzheimer's disease: hippocampal shape and volume predict dementia onset in the elderly. Neuroimage 2005;25:783–792.

20 Hsu YY, Schuff N, Du AT, Mark K, Zhu X, Hardin D, et al: Comparison of automated and manual MRI volumetry of hippocampus in normal aging and dementia. J Magn Reson Imaging 2002;16:305–310.

21 Colliot O, Chetelat G, Chupin M, Desgranges B, Magnin B, Benali H, et al: Discrimination between Alzheimer disease, mild cognitive impairment, and normal aging by using automated segmentation of the hippocampus. Radiology 2008;248:194–201.

22 Calvini P, Chincarini A, Gemme G, Penco MA, Squarcia S, Nobili F, et al: Automatic analysis of medial temporal lobe atrophy from structural MRIs for the early assessment of Alzheimer disease. Med Phys 2009;36:3737–3747.

23 Krasuski JS, Alexander GE, Horwitz B, Daly EM, Murphy DGM, Rapoport SI, et al: Volumes of medial temporal lobe structures in patients with Alzheimer's disease and mild cognitive impairment (and in healthy controls). Biol Psychiatry 1998;43: 60–68.

24 Pennanen C, Kivipelto M, Tuomainen S, Hartikainen P, Hanninen T, Laakso MP, et al: Hippocampus and entorhinal cortex in mild cognitive impairment and early AD. Neurobiol Aging 2004;25:303–310.
25 Teipel SJ, Pruessner JC, Faltraco F, Born C, Rocha-Unold M, Evans A, et al: Comprehensive dissection of the medial temporal lobe in AD: measurement of hippocampus, amygdala, entorhinal, perirhinal and parahippocampal cortices using MRI. J Neurol 2006;253:794–800.
26 Xu Y, Jack CR Jr, O'Brien PC, Kokmen E, Smith GE, Ivnik RJ, et al: Usefulness of MRI measures of entorhinal cortex versus hippocampus in AD. Neurology 2000;54:1760–1767.
27 Du AT, Schuff N, Amend D, Laakso MP, Hsu YY, Jagust WJ, et al: Magnetic resonance imaging of the entorhinal cortex and hippocampus in mild cognitive impairment and Alzheimer's disease. J Neurol Neurosurg Psychiatry 2001;71:441–447.
28 Risacher SL, Saykin AJ, West JD, Shen L, Firpi HA, McDonald BC: Baseline MRI predictors of conversion from MCI to probable AD in the ADNI cohort. Curr Alzheimer Res 2009;6:347–361.
29 Teipel SJ, Flatz WH, Heinsen H, Bokde AL, Schoenberg SO, Stockel S, et al: Measurement of basal forebrain atrophy in Alzheimer's disease using MRI. Brain 2005;128:2626–2644.
30 Muth K, Schonmeyer R, Matura S, Haenschel C, Schroder J, Pantel J: Mild Cognitive Impairment in the elderly is associated with volume loss of the cholinergic basal forebrain region. Biol Psychiatry 2010; 67:588–591.
31 Grothe M, Zaborszky L, Atienza M, Gil-Neciga E, Rodriguez-Romero R, Teipel SJ, et al: Reduction of basal forebrain cholinergic system parallels cognitive impairment in patients at high risk of developing Alzheimer's disease. Cereb Cortex 2010;20: 1685–1695.
32 Teipel SJ, Meindl T, Grinberg L, Grothe M, Cantero JL, Reiser MF, et al: The cholinergic system in mild cognitive impairment and Alzheimer's disease: an in vivo MRI and DTI study. Hum Brain Mapp 2011; 32:1349–1362.
33 Ashburner J, Friston KJ: Voxel-based morphometry – the methods. Neuroimage 2000;11:805–821.
34 Good CD, Johnsrude IS, Ashburner J, Henson RN, Friston KJ, Frackowiak RS: A voxel-based morphometric study of ageing in 465 normal adult human brains. Neuroimage 2001;14:21–36.
35 Baron JC, Chetelat G, Desgranges B, Perchey G, Landeau B, de la Sayette V, et al: In vivo mapping of gray matter loss with voxel-based morphometry in mild Alzheimer's disease. Neuroimage 2001;14: 298–309.
36 Busatto GF, Garrido GE, Almeida OP, Castro CC, Camargo CH, Cid CG, et al: A voxel-based morphometry study of temporal lobe gray matter reductions in Alzheimer's disease. Neurobiol Aging 2003; 24:221–231.
37 Chetelat G, Desgranges B, De La Sayette V, Viader F, Eustache F, Baron JC: Mapping gray matter loss with voxel-based morphometry in mild cognitive impairment. Neuroreport 2002;13:1939–1943.
38 Pennanen C, Testa C, Laakso MP, Hallikainen M, Helkala EL, Hanninen T, et al: A voxel-based morphometry study on mild cognitive impairment. J Neurol Neurosurg Psychiatry 2005;76:11–14.
39 Teipel SJ, Alexander GE, Schapiro MB, Moller HJ, Rapoport SI, Hampel H: Age-related cortical gray matter reductions in non-demented Down's syndrome adults determined by MRI with voxel-based morphometry. Brain 2004;127:811–824.
40 Teipel SJ, Hampel H: Neuroanatomy of Down syndrome in vivo: a model of preclinical Alzheimer's disease. Behav Genet 2006;36:405–415.
41 Karas GB, Scheltens P, Rombouts SA, Visser PJ, van Schijndel RA, Fox NC, et al: Global and local gray matter loss in mild cognitive impairment and Alzheimer's disease. Neuroimage 2004;23:708–716.
42 Smith CD, Chebrolu H, Wekstein DR, Schmitt FA, Jicha GA, Cooper G, et al: Brain structural alterations before mild cognitive impairment. Neurology 2007;68:1268–1273.
43 Di Paola M, Macaluso E, Carlesimo GA, Tomaiuolo F, Worsley KJ, Fadda L, et al: Episodic memory impairment in patients with Alzheimer's disease is correlated with entorhinal cortex atrophy. A voxel-based morphometry study. J Neurol 2007;254: 774–781.
44 Grossman M, McMillan C, Moore P, Ding L, Glosser G, Work M, et al: What's in a name: voxel-based morphometric analyses of MRI and naming difficulty in Alzheimer's disease, frontotemporal dementia and corticobasal degeneration. Brain 2004;127: 628–649.
45 Josephs KA, Whitwell JL, Duffy JR, Vanvoorst WA, Strand EA, Hu WT, et al: Progressive aphasia secondary to Alzheimer disease vs FTLD pathology. Neurology 2008;70:25–34.
46 Gee J, Ding L, Xie Z, Lin M, DeVita C, Grossman M: Alzheimer's disease and frontotemporal dementia exhibit distinct atrophy-behavior correlates: a computer-assisted imaging study. Acad Radiol 2003; 10:1392–1401.
47 Thompson PM, Hayashi KM, de Zubicaray G, Janke AL, Rose SE, Semple J, et al: Dynamics of gray matter loss in Alzheimer's disease. J Neurosci 2003;23: 994–1005.

48 Thompson PM, Hayashi KM, Dutton RA, Chiang MC, Leow AD, Sowell ER, et al: Tracking Alzheimer's disease. Ann NY Acad Sci 2007;1097:183–214.

49 Davatzikos C, Fan Y, Wu X, Shen D, Resnick SM: Detection of prodromal Alzheimer's disease via pattern classification of MRI. Neurobiol Aging 2006.

50 Fan Y, Resnick SM, Wu X, Davatzikos C: Structural and functional biomarkers of prodromal Alzheimer's disease: a high-dimensional pattern classification study. Neuroimage 2008;41:277–285.

51 Fan Y, Shen D, Davatzikos C: Classification of structural images via high-dimensional image warping, robust feature extraction, and SVM. Med. Image Comput. Comput. Assist. Interv. Int. Conf. Med. Image Comput. Comput. Assist. Interv 2005;8:1–8.

52 Plant C, Teipel SJ, Oswald A, Bohm C, Meindl T, Mourao-Miranda J, et al: Automated detection of brain atrophy patterns based on MRI for the prediction of Alzheimer's disease. Neuroimage 2010;50: 162–174.

53 Teipel SJ, Born C, Ewers M, Bokde AL, Reiser MF, Moller HJ, et al: Multivariate deformation-based analysis of brain atrophy to predict Alzheimer's disease in mild cognitive impairment. Neuroimage 2007;38:13–24.

54 Kloppel S, Stonnington CM, Chu C, Draganski B, Scahill RI, Rohrer JD, et al: Automatic classification of MR scans in Alzheimer's disease. Brain 2008;131: 681–689.

55 Acosta O, Bourgeat P, Zuluaga MA, Fripp J, Salvado O, Ourselin S: Automated voxel-based 3D cortical thickness measurement in a combined Lagrangian-Eulerian PDE approach using partial volume maps. Med Image Anal 2009;13:730–743.

56 Lerch J, Pruessner JC, Evans AC, Zijdenbos A, Teipel S, Buerger K, et al: Cortical thickness in Alzheimer's disease. Neuroimage Suppl Human Brain Mapping Conference, Sendai 2002.

57 Desikan RS, Cabral HJ, Hess CP, Dillon WP, Glastonbury CM, Weiner MW, et al: Automated MRI measures identify individuals with mild cognitive impairment and Alzheimer's disease. Brain 2009;132:2048–2057.

58 Lerch JP, Pruessner J, Zijdenbos AP, Collins DL, Teipel SJ, Hampel H, et al: Automated cortical thickness measurements from MRI can accurately separate Alzheimer's patients from normal elderly controls. Neurobiol Aging 2008;29:23–30.

59 Oliveira PP Jr, Nitrini R, Busatto G, Buchpiguel C, Sato JR, Amaro E Jr: Use of SVM methods with surface-based cortical and volumetric subcortical measurements to detect Alzheimer's disease. J Alzheimers Dis 2010;19:1263–1272.

60 Bandettini PA: What's new in neuroimaging methods? Ann NY Acad Sci 2009;1156:260–293.

61 Mori S, Zhang J: Principles of diffusion tensor imaging and its applications to basic neuroscience research. Neuron 2006;51:527–539.

62 Sjobeck M, Haglund M, Englund E: White matter mapping in Alzheimer's disease: a neuropathological study. Neurobiol Aging 2006;27:673–680.

63 Sjobeck M, Haglund M, Englund E: Decreasing myelin density reflected increasing white matter pathology in Alzheimer's disease – a neuropathological study. Int J Geriatr Psychiatry 2005;20: 919–926.

64 Gouw AA, Seewann A, Vrenken H, van der Flier WM, Rozemuller JM, Barkhof F, et al: Heterogeneity of white matter hyperintensities in Alzheimer's disease: post-mortem quantitative MRI and neuropathology. Brain 2008;131:3286–3298.

65 Bartzokis G: Age-related myelin breakdown: a developmental model of cognitive decline and Alzheimer's disease. Neurobiol Aging 2004;25:5–18.

66 Bartzokis G, Cummings JL, Sultzer D, Henderson VW, Nuechterlein KH, Mintz J: White matter structural integrity in healthy aging adults and patients with Alzheimer disease: a magnetic resonance imaging study. Arch Neurol 2003;60:393–398.

67 Sherin J, Bartzokis, G: Human brain myelination trajectories across the fife span: implications for CNS function and dysfunction; in Masoro EJ, Austad SN (eds): Handbook of the Biology of Aging, ed 7. London, Elsevier, 2011.

68 De Reuck J, Deramecourt V, Cordonnier C, Leys D, Maurage CA, Pasquier F: The impact of cerebral amyloid angiopathy on the occurrence of cerebrovascular lesions in demented patients with Alzheimer features: a neuropathological study. Eur J Neurol 2011;18:913–918.

69 Avants BB, Cook PA, Ungar L, Gee JC, Grossman M: Dementia induces correlated reductions in white matter integrity and cortical thickness: a multivariate neuroimaging study with sparse canonical correlation analysis. Neuroimage 2010;50:1004–1016.

70 Basser PJ, Mattiello J, LeBihan D: MR diffusion tensor spectroscopy and imaging. Biophys J 1994;66: 259–267.

71 Pierpaoli C, Jezzard P, Basser PJ, Barnett A, Di Chiro G: Diffusion tensor MR imaging of the human brain. Radiology 1996;201:637–648.

72 Le Bihan D, Mangin JF, Poupon C, Clark CA, Pappata S, Molko N, et al: Diffusion tensor imaging: concepts and applications. J Magn Reson Imaging 2001;13:534–546.

73 Yesavage JA, Mumenthaler MS, Taylor JL, Friedman L, O'Hara R, Sheikh J, et al: Donepezil and flight simulator performance: effects on retention of complex skills. Neurology 2002;59:123–125.

74 Pierpaoli C, Basser PJ: Toward a quantitative assessment of diffusion anisotropy. Magn Reson Med 1996;36:893–906.

75 Mori S, Crain BJ, Chacko VP, van Zijl PC: Three-dimensional tracking of axonal projections in the brain by magnetic resonance imaging. Ann Neurol 1999;45:265–269.

76 Hanyu H, Sakurai H, Iwamoto T, Takasaki M, Shindo H, Abe K: Diffusion-weighted MR imaging of the hippocampus and temporal white matter in Alzheimer's disease. J Neurol Sci 1998;156:195–200.

77 Bozzali M, Franceschi M, Falini A, Pontesilli S, Cercignani M, Magnani G, et al: Quantification of tissue damage in AD using diffusion tensor and magnetization transfer MRI. Neurology 2001;57: 1135–1137.

78 Zhang Y, Schuff N, Jahng GH, Bayne W, Mori S, Schad L, et al: Diffusion tensor imaging of cingulum fibers in mild cognitive impairment and Alzheimer disease. Neurology 2007;68:13–19.

79 Bozzali M, Falini A, Franceschi M, Cercignani M, Zuffi M, Scotti G, et al: White matter damage in Alzheimer's disease assessed in vivo using diffusion tensor magnetic resonance imaging. J Neurol Neurosurg Psychiatry 2002;72:742–726.

80 Yoshiura T, Mihara F, Ogomori K, Tanaka A, Kaneko K, Masuda K: Diffusion tensor in posterior cingulate gyrus: correlation with cognitive decline in Alzheimer's disease. Neuroreport 2002;13:2299–2302.

81 Takahashi S, Yonezawa H, Takahashi J, Kudo M, Inoue T, Tohgi H: Selective reduction of diffusion anisotropy in white matter of Alzheimer disease brains measured by 3.0 Tesla magnetic resonance imaging. Neurosci Lett 2002;332:45–48.

82 Rose SE, Chen F, Chalk JB, Zelaya FO, Strugnell WE, Benson M, et al: Loss of connectivity in Alzheimer's disease: an evaluation of white matter tract integrity with colour-coded MR diffusion tensor imaging. J Neurol Neurosurg Psychiatry 2000;69: 528–530.

83 Fellgiebel A, Wille P, Muller MJ, Winterer G, Scheurich A, Vucurevic G, et al: Ultrastructural hippocampal and white matter alterations in mild cognitive impairment: a diffusion tensor imaging study. Dement Geriatr Cogn Disord 2004;18:101–108.

84 Fellgiebel A, Muller MJ, Wille P, Dellani PR, Scheurich A, Schmidt LG, et al: Color-coded diffusion-tensor imaging of posterior cingulate fiber tracts in mild cognitive impairment. Neurobiol Aging 2005;26:1193–1198.

85 Kantarci K, Jack CR Jr, Xu YC, Campeau NG, O'Brien PC, Smith GE, et al: Mild cognitive impairment and Alzheimer disease: regional diffusivity of water. Radiology 2001;219:101–107.

86 Kalus P, Slotboom J, Gallinat J, Mahlberg R, Cattapan-Ludewig K, Wiest R, et al: Examining the gateway to the limbic system with diffusion tensor imaging: the perforant pathway in dementia. Neuroimage 2006;30:713–720.

87 Teipel SJ, Stahl R, Dietrich O, Schoenberg SO, Perneczky R, Bokde AL, et al: Multivariate network analysis of fiber tract integrity in Alzheimer's disease. Neuroimage 2007;34:985–995.

88 Xie S, Xiao JX, Gong GL, Zang YF, Wang YH, Wu HK, et al: Voxel-based detection of white matter abnormalities in mild Alzheimer disease. Neurology 2006;66:1845–1849.

89 Sexton CE, Kalu UG, Filippini N, Mackay CE, Ebmeier KP: A meta-analysis of diffusion tensor imaging in mild cognitive impairment and Alzheimer's disease. Neurobiol Aging.

90 Cherubini A, Peran P, Spoletini I, Di Paola M, Di Iulio F, Hagberg GE, et al: Combined volumetry and DTI in subcortical structures of mild cognitive impairment and Alzheimer's disease patients. J Alzheimers Dis 2010;19:1273–1282.

91 Canu E, McLaren DG, Fitzgerald ME, Bendlin BB, Zoccatelli G, Alessandrini F, et al: Microstructural diffusion changes are independent of macrostructural volume loss in moderate to severe Alzheimer's disease. J Alzheimer Dis 2010;19:963–976.

92 Fellgiebel A, Dellani PR, Greverus D, Scheurich A, Stoeter P, Muller MJ: Predicting conversion to dementia in mild cognitive impairment by volumetric and diffusivity measurements of the hippocampus. Psychiatry Res 2006;146:283–287.

93 Kantarci K, Petersen RC, Boeve BF, Knopman DS, Weigand SD, O'Brien PC, et al: DWI predicts future progression to Alzheimer disease in amnestic mild cognitive impairment. Neurology 2005;64:902–904.

94 Friese U, Meindl T, Herpertz SC, Reiser MF, Hampel H, Teipel SJ: Diagnostic utility of novel MRI-based biomarkers for Alzheimer's disease: diffusion tensor imaging and deformation-based morphometry. J Alzheimers Dis 2010;20:477–490.

95 Lo CY, Wang PN, Chou KH, Wang J, He Y, Lin CP: Diffusion tensor tractography reveals abnormal topological organization in structural cortical networks in Alzheimer's disease. J Neurosci 2010;30: 16876–16885.

96 Yakushev I, Gerhard A, Muller MJ, Lorscheider M, Buchholz HG, Schermuly I, et al: Relationships between hippocampal microstructure, metabolism, and function in early Alzheimer's disease. Brain Struct Funct 2011;216:219–226.

97 Teipel SJ, Bokde AL, Meindl T, Amaro E Jr, Soldner J, Reiser MF, et al: White matter microstructure underlying default mode network connectivity in the human brain. Neuroimage 2010;49:2021–2032.

98 Teipel SJ, Meindl T, Wagner M, Kohl T, Burger K, Reiser MF, et al: White matter microstructure in relation to education in aging and Alzheimer's disease. J Alzheimers Dis 2009;17:571–583.
99 Brewer JB, Zhao Z, Desmond JE, Glover GH, Gabrieli JD: Making memories: brain activity that predicts how well visual experience will be remembered. Science 1998;281:1185–1187.
100 Wagner AD, Schacter DL, Rotte M, Koutstaal W, Maril A, Dale AM, et al: Building memories: remembering and forgetting of verbal experiences as predicted by brain activity. Science 1998;281: 1188–1191.
101 Sperling R, Chua E, Cocchiarella A, Rand-Giovannetti E, Poldrack R, Schacter DL, et al: Putting names to faces: successful encoding of associative memories activates the anterior hippocampal formation. Neuroimage 2003;20:1400–1410.
102 Miller SL, Celone K, DePeau K, Diamond E, Dickerson BC, Rentz D, et al: Age-related memory impairment associated with loss of parietal deactivation but preserved hippocampal activation. Proc Natl Acad Sci USA 2008;105:2181–2186.
103 Ogawa S, Lee TM, Nayak AS, Glynn P: Oxygenation-sensitive contrast in magnetic resonance image of rodent brain at high magnetic fields. Magn Reson Med 1990;14:68–78.
104 Kwong KK, Belliveau JW, Chesler DA, Goldberg IE, Weisskoff RM, Poncelet BP, et al: Dynamic magnetic resonance imaging of human brain activity during primary sensory stimulation. Proc Natl Acad Sci USA 1992;89:5675–5679.
105 Logothetis NK, Pauls J, Augath M, Trinath T, Oeltermann A: Neurophysiological investigation of the basis of the fMRI signal. Nature 2001;412:150–157.
106 Kato T, Knopman D, Liu H: Dissociation of regional activation in mild AD during visual encoding: a functional MRI study. Neurology 2001;57:812–816.
107 Rombouts SA, Barkhof F, Veltman DJ, Machielsen WC, Witter MP, Bierlaagh MA, et al: Functional MR imaging in Alzheimer's disease during memory encoding. AJNR Am J Neuroradiol 2000;21:1869–1875.
108 Small SA, Perera GM, DeLaPaz R, Mayeux R, Stern Y: Differential regional dysfunction of the hippocampal formation among elderly with memory decline and Alzheimer's disease. Ann Neurol 1999; 45:466–472.
109 Gron G, Bittner D, Schmitz B, Wunderlich AP, Riepe MW: Subjective memory complaints: objective neural markers in patients with Alzheimer's disease and major depressive disorder. Ann Neurol 2002;51:491–498.
110 Sperling R, Bates J, Chua E, Cocchiarella A, Schacter DL, Rosen B, et al: fMRI studies of associative encoding in young and elderly controls and mild AD patients. J Neurol Neurosurg Psychiatry 2003; 74:44–50.
111 Machulda MM, Ward HA, Borowski B, Gunter JL, Cha RH, O'Brien PC, et al: Comparison of memory fMRI response among normal, MCI, and Alzheimer's patients. Neurology 2003;61:500–506.
112 Golby A, Silverberg G, Race E, Gabrieli S, O'Shea J, Knierim K, et al: Memory encoding in Alzheimer's disease: an fMRI study of explicit and implicit memory. Brain 2005;128:773–787.
113 Remy F, Mirrashed F, Campbell B, Richter W: Mental calculation impairment in Alzheimer's disease: a functional magnetic resonance imaging study. Neurosci Lett 2004;358:25–28.
114 Hamalainen A, Pihlajamaki M, Tanila H, Hanninen T, Niskanen E, Tervo S, et al: Increased fMRI responses during encoding in mild cognitive impairment. Neurobiol Aging 2007;28:1889–1903.
115 Schwindt GC, Black SE: Functional imaging studies of episodic memory in Alzheimer's disease: a quantitative meta-analysis. Neuroimage 2009;45:181–190.
116 Grady CL, McIntosh AR, Beig S, Keightley ML, Burian H, Black SE: Evidence from functional neuroimaging of a compensatory prefrontal network in Alzheimer's disease. J Neurosci 2003;23:986–993.
117 Small SA, Perera GM, DeLaPaz R, Mayeux R, Stern Y: Differential regional dysfunction of the hippocampal formation among elderly with memory decline and Alzheimer's disease. Ann Neurol 1999; 45:466–472.
118 Johnson SC, Schmitz TW, Moritz CH, Meyerand ME, Rowley HA, Alexander AL, et al: Activation of brain regions vulnerable to Alzheimer's disease: the effect of mild cognitive impairment. Neurobiol Aging 2006;27:1604–1612.
119 Petrella JR, Krishnan S, Slavin M, Tran T-T, Murty L, Doraiswamy P: Mild cognitive impairment: evaluation with 4-T functional MR imaging. Radiology 2006;240:177–186.
120 Smith CD, Andersen AH, Kryscio RJ, Schmitt FA, Kindy MS, Blonder LX, et al: Altered brain activation in cognitively intact individuals at high risk for Alzheimer's disease. Neurology 1999;53:1391–1396.
121 Lind J, Ingvar M, Persson J, Sleegers K, Van Broeckhoven C, Adolfsson R, et al: Parietal cortex activation predicts memory decline in apolipoprotein E ε4 carriers. Neuroreport 2006;17:1683–1686.
122 Lind J, Larsson A, Persson J, Ingvar M, Nilsson LG, Backman L, et al: Reduced hippocampal volume in non-demented carriers of the apolipoprotein E ε4: relation to chronological age and recognition memory. Neurosci Lett 2006;396:23–27.

123 Trivedi MA, Schmitz TW, Ries ML, Torgerson BM, Sager MA, Hermann BP, et al: Reduced hippocampal activation during episodic encoding in middle-aged individuals at genetic risk of Alzheimer's disease: a cross-sectional study. BMC Med 2006;4:1.

124 Borghesani PR, Johnson LC, Shelton AL, Peskind ER, Aylward EH, Schellenberg GD, et al: Altered medial temporal lobe responses during visuospatial encoding in healthy APOE*4 carriers. Neurobiol Aging 2007;29:981–991.

125 Mondadori CR, de Quervain DJ, Buchmann A, Mustovic H, Wollmer MA, Schmidt CF, et al: Better memory and neural efficiency in young apolipoprotein E ε4 carriers. Cereb Cortex 2007;17:1934–1947.

126 Dickerson BC, Salat DH, Bates JF, Atiya M, Killiany RJ, Greve DN, et al: Medial temporal lobe function and structure in mild cognitive impairment. Ann Neurol 2004;56:27–35.

127 Dickerson BC, Salat DH, Greve D, Chua E, Rand-Giovannetti E, Rentz D, et al: Increased hippocampal activation in mild cognitive impairment compared to normal aging and AD. Neurology 2005;65:404–411.

128 Hamalainen A, Pihlajamaki M, Tanila H, Hanninen T, Niskanen E, Tervo S, et al: Increased fMRI responses during encoding in mild cognitive impairment. Neurobiol Aging 2007;28:1889–1903.

129 Celone KA, Calhoun VD, Dickerson BC, Atri A, Chua EF, Miller SL, et al: Alterations in memory networks in mild cognitive impairment and Alzheimer's disease: an independent component analysis. J Neurosci 2006;26:10222–10231.

130 Kircher TT, Weis S, Freymann K, Erb M, Jessen F, Grodd W, et al: Hippocampal activation in patients with mild cognitive impairment is necessary for successful memory encoding. J Neurol Neurosurg Psychiatry 2007;78:812–818.

131 Heun R, Freymann K, Erb M, Leube DT, Jessen F, Kircher TT, et al: Mild cognitive impairment and actual retrieval performance affect cerebral activation in the elderly. Neurobiol Aging 2007;28:404–413.

132 Bookheimer SY, Strojwas MH, Cohen MS, Saunders AM, Pericak-Vance MA, Mazziotta JC, et al: Patterns of brain activation in people at risk for Alzheimer's disease. N Engl J Med 2000;343:450–456.

133 Smith CD, Andersen AH, Kryscio RJ, Schmitt FA, Kindy MS, Blonder LX, et al: Women at risk for AD show increased parietal activation during a fluency task. Neurology 2002;58:1197–1202.

134 Bondi MW, Houston WS, Eyler LT, Brown GG: fMRI evidence of compensatory mechanisms in older adults at genetic risk for Alzheimer disease. Neurology 2005;64:501–508.

135 Fleisher AS, Houston WS, Eyler LT, Frye S, Jenkins C, Thal LJ, et al: Identification of Alzheimer disease risk by functional magnetic resonance imaging. Arch Neurol 2005;62:1881–1888.

136 Wishart HA, Saykin AJ, McDonald BC, Mamourian AC, Flashman LA, Schuschu KR, et al: Brain activation patterns associated with working memory in relapsing-remitting MS. Neurology 2004;62:234–238.

137 Han SD, Houston WS, Jak AJ, Eyler LT, Nagel BJ, Fleisher AS, et al: Verbal paired-associate learning by APOE genotype in non-demented older adults: fMRI evidence of a right hemispheric compensatory response. Neurobiol Aging 2007;28:238–247.

138 Filippini N, MacIntosh BJ, Hough MG, Goodwin GM, Frisoni GB, Smith SM, et al: Distinct patterns of brain activity in young carriers of the APOE ε4 allele. Proc Natl Acad Sci USA 2009;106:7209–7214.

139 Dickerson BC, Sperling RA: Functional abnormalities of the medial temporal lobe memory system in mild cognitive impairment and Alzheimer's disease: insights from functional MRI studies. Neuropsychologia 2008;46:1624–1635.

140 Sperling RA, Laviolette PS, O'Keefe K, O'Brien J, Rentz DM, Pihlajamaki M, et al: Amyloid deposition is associated with impaired default network function in older persons without dementia. Neuron 2009;63:178–188.

141 Miller SL, Fenstermacher E, Bates J, Blacker D, Sperling RA, Dickerson BC: Hippocampal activation in adults with mild cognitive impairment predicts subsequent cognitive decline. J Neurol Neurosurg Psychiatry 2008;79:630–635.

142 O'Brien JL, O'Keefe KM, LaViolette PS, DeLuca AN, Blacker D, Dickerson BC, et al: Longitudinal fMRI in elderly reveals loss of hippocampal activation with clinical decline. Neurology 2010;74:1969–1976.

143 DeKosky ST, Ikonomovic MD, Styren SD, Beckett L, Wisniewski S, Bennett DA, et al: Upregulation of choline acetyltransferase activity in hippocampus and frontal cortex of elderly subjects with mild cognitive impairment. Ann Neurol 2002;51:145–155.

144 Hashimoto S, Murakami M, Kanaseki T, Kobayashi S, Matsuki M, Shimono M, et al: Morphological and functional changes in cell junctions during secretory stimulation in the perfused rat submandibular gland. Eur J Morphol 2003;41:35–39.

145 Stern EA, Bacskai BJ, Hickey GA, Attenello FJ, Lombardo JA, Hyman BT: Cortical synaptic integration in vivo is disrupted by amyloid-β plaques. J Neurosci 2004;24:4535–4540.

146 Busche MA, Eichhoff G, Adelsberger H, Abramowski D, Wiederhold KH, Haass C, et al: Clusters of hyperactive neurons near amyloid plaques in a mouse model of Alzheimer's disease. Science 2008; 321:1686–1689.

147 Palop JJ, Chin J, Roberson ED, Wang J, Thwin MT, Bien-Ly N, et al: Aberrant excitatory neuronal activity and compensatory remodeling of inhibitory hippocampal circuits in mouse models of Alzheimer's disease. Neuron 2007;55:697–711.

148 Raichle ME, MacLeod AM, Snyder AZ, Powers WJ, Gusnard DA, Shulman GL: A default mode of brain function. Proc Natl Acad Sci USA 2001;98:676–682.

149 Buckner RL, Andrews-Hanna JR, Schacter DL: The brain's default network: anatomy, function, and relevance to disease. Ann NY Acad Sci 2008;1124:1–38.

150 Lustig C, Buckner RL: Preserved neural correlates of priming in old age and dementia. Neuron 2004; 42:865–875.

151 Petrella JR, Prince SE, Wang L, Hellegers C, Doraiswamy PM: Prognostic value of posteromedial cortex deactivation in mild cognitive impairment. PLoS One 2007;2:e1104.

152 Pihlajamaki M, Depeau KM, Blacker D, Sperling RA: Impaired medial temporal repetition suppression is related to failure of parietal deactivation in Alzheimer disease. Am J Geriatr Psychiatry 2008; 16:283–292.

153 Pihlajamaki M, O'Keefe K, Bertram L, Tanzi R, Dickerson B, Blacker D, et al: Evidence of altered posteromedial cortical fMRI activity in subjects at risk for Alzheimer disease. Alzheimer Dis Assoc Disord 2010;24:28–36.

154 Vannini P, Hedden T, Becker JA, Sullivan C, Putcha D, Rentz D, et al: Age and amyloid-related alterations in default network habituation to stimulus repetition. Neurobiol Aging 2011 (E-pub ahead of print).

155 Greicius MD, Srivastava G, Reiss AL, Menon V: Default-mode network activity distinguishes Alzheimer's disease from healthy aging: evidence from functional MRI. Proc Natl Acad Sci USA 2004;101:4637–4642.

156 Petrella JR, Sheldon FC, Prince SE, Calhoun VD, Doraiswamy PM: Default mode network connectivity in stable vs. progressive mild cognitive impairment. Neurology 2011;76:511–517.

157 Klunk WE, Engler H, Nordberg A, Wang Y, Blomqvist G, Holt DP, et al: Imaging brain amyloid in Alzheimer's disease with Pittsburgh Compound B. Ann Neurol 2004;55:306–319.

158 Buckner RL, Snyder AZ, Shannon BJ, LaRossa G, Sachs R, Fotenos AF, et al: Molecular, structural, and functional characterization of Alzheimer's disease: evidence for a relationship between default activity, amyloid, and memory. J Neurosci 2005;25:7709–7717.

159 Buckner RL, Sepulcre J, Talukdar T, Krienen FM, Liu H, Hedden T, et al: Cortical hubs revealed by intrinsic functional connectivity: mapping, assessment of stability, and relation to Alzheimer's disease. J Neurosci 2009;29:1860–1873.

160 Hedden T, Van Dijk KR, Becker JA, Mehta A, Sperling RA, Johnson KA, et al: Disruption of functional connectivity in clinically normal older adults harboring amyloid burden. J Neurosci 2009;29: 12686–12694.

161 Sheline YI, Raichle ME, Snyder AZ, Morris JC, Head D, Wang S, et al: Amyloid plaques disrupt resting state default mode network connectivity in cognitively normal elderly. Biol Psychiatry 2010;67:584–587.

162 Thiel CM, Henson RN, Morris JS, Friston KJ, Dolan RJ: Pharmacological modulation of behavioral and neuronal correlates of repetition priming. J Neurosci 2001;21:6846–6852.

163 Sperling RA, Greve D, Dale A, Killiany R, Rosen B, Holmes J, et al: fMRI detection of pharmacologically induced memory impairment. Proc Natl Acad Sci USA 2002;99:455–460.

164 Kukolja J, Thiel CM, Fink GR: Cholinergic stimulation enhances neural activity associated with encoding but reduces neural activity associated with retrieval in humans. J Neurosci 2009;29:8119–8128.

165 Rombouts SA, Barkhof F, Van Meel CS, Scheltens P: Alterations in brain activation during cholinergic enhancement with rivastigmine in Alzheimer's disease. J Neurol Neurosurg Psychiatry 2002;73: 665–671.

166 Saykin AJ, Wishart HA, Rabin LA, Flashman LA, McHugh TL, Mamourian AC, et al: Cholinergic enhancement of frontal lobe activity in mild cognitive impairment. Brain 2004;127:1574–1583.

167 Goekoop R, Rombouts SA, Jonker C, Hibbel A, Knol DL, Truyen L, et al: Challenging the cholinergic system in mild cognitive impairment: a pharmacological fMRI study. Neuroimage 2004;23:1450–1459.

168 Shanks MF, McGeown WJ, Forbes-McKay KE, Waiter GD, Ries M, Venneri A: Regional brain activity after prolonged cholinergic enhancement in early Alzheimer's disease. Magn Reson Imaging 2007;25:848–859.

169 Bokde AL, Karmann M, Teipel SJ, Born C, Lieb M, Reiser MF, et al: Decreased activation along the dorsal visual pathway after a 3-month treatment with galantamine in mild Alzheimer disease: a functional magnetic resonance imaging study. J Clin Psychopharmacol 2009;29:147–156.

170 Venneri A, McGeown WJ, Shanks MF: Responders to ChEI treatment of Alzheimer's disease show restitution of normal regional cortical activation. Curr Alzheimer Res 2009;6:97–111.

171 Hampstead BM, Stringer AY, Stilla RF, Deshpande G, Hu X, Moore AB, et al: Activation and effective connectivity changes following explicit-memory training for face-name pairs in patients with mild cognitive impairment: a pilot study. Neurorehabil Neural Repair 2011;25:210–222.

172 Sperling R, Greve D, Dale A, Killiany R, Holmes J, Rosas HD, et al: Functional MRI detection of pharmacologically induced memory impairment. Proc Natl Acad Sci USA 2002;99:455–460.

173 Meindl T, Teipel S, Elmouden R, Mueller S, Koch W, Dietrich O, et al: Test-retest reproducibility of the default-mode network in healthy individuals. Hum Brain Mapp 2010;31:237–246.

174 Zuo XN, Kelly C, Adelstein JS, Klein DF, Castellanos FX, Milham MP: Reliable intrinsic connectivity networks: test-retest evaluation using independent component analysis and dual regression approach. Neuroimage 2010;49:2163–177.

175 Clement F, Belleville S: Test-retest reliability of fMRI verbal episodic memory paradigms in healthy older adults and in persons with mild cognitive impairment. Hum Brain Mapp 2009;30:4033–4047.

176 Putcha D, O'Keefe K, Laviolette P, O'Brien J, Greve D, Rentz DM, et al: Reliability of functional magnetic resonance imaging associative encoding memory paradigms in non-demented elderly adults. Hum Brain Mapp 2011;32:2027–2044.

177 Jack CR Jr, Knopman DS, Jagust WJ, Shaw LM, Aisen PS, Weiner MW, et al: Hypothetical model of dynamic biomarkers of the Alzheimer's pathological cascade. Lancet Neurol 2010;9:119–128.

178 Calhoun VD, Adali T, McGinty VB, Pekar JJ, Watson TD, Pearlson GD: fMRI activation in a visual-perception task: Network of areas detected using the general linear model and independent components analysis. Neuroimage 2001;14:1080–1088.

179 Van de Ven VG, Formisano E, Prvulovic D, Roeder CH, Linden DE: Functional connectivity as revealed by spatial independent component analysis of fMRI measurements during rest. Hum Brain Mapp 2004;22:165–178.

180 Beckmann CF, DeLuca M, Devlin JT, Smith SM: Investigations into resting-state connectivity using independent component analysis. Philos Trans R Soc Lond B Biol Sci 2005;360:1001–1013.

181 Hedden T, Van Dijk KRA, Becker JA, Mehta A, Sperling RA, Johnson KA, et al: Disruption of functional connectivity in clinically normal older adults harboring amyloid burden. J Neurosci 2009;29:12686–12694.

182 Sheline YI, Morris JC, Snyder AZ, Price JL, Yan ZZ, D'Angelo G, et al: APOE4 allele disrupts resting state fMRI connectivity in the absence of amyloid plaques or decreased CSF $A\beta_{42}$. J Neurosci 2010;30:17035–17040.

183 Li SJ, Li Z, Wu GH, Zhang MJ, Franczak M, Antuono PG: Alzheimer disease: Evaluation of a functional MR imaging index as a marker. Radiology 2002;225:253–259.

184 Greicius MD, Srivastava G, Reiss AL, Menon V: Default-mode network activity distinguishes Alzheimer's disease from healthy aging: evidence from functional MRI. Proc Natl Acad Sci USA 2004;101:4637–4642.

185 Allen G, Barnard H, McColl R, Hester AL, Fields JA, Weiner MF, et al: Reduced hippocampal functional connectivity in Alzheimer disease. Arch Neurol 2007;64:1482–1487.

186 Wang L, Zang YF, He Y, Liang M, Zhang XQ, Tian LX, et al: Changes in hippocampal connectivity in the early stages of Alzheimer's disease: evidence from resting state fMRI. Neuroimage 2006;31:496–504.

187 Greicius MD, Supekar K, Menon V, Dougherty RF: Resting-state functional connectivity reflects structural connectivity in the default mode network. Cereb Cortex 2009;19:72–78.

188 Koch W, Teipel S, Mueller S, Benninghoff J, Wagner M, Bokde AL, et al: Diagnostic power of default mode network resting state fMRI in the detection of Alzheimer's disease. Neurobiol Aging 2011 (E-pub ahead of print).

189 Sorg C, Riedl V, Muhlau M, Calhoun VD, Eichele T, Laer L, et al: Selective changes of resting-state networks in individuals at risk for Alzheimer's disease. Proc Natl Acad Sci USA 2007;104:18760–18765.

190 Grothe M, Zaborszky L, Atienza M, Gil-Neciga E, Rodriguez-Romero R, Teipel SJ, et al: Reduction of basal forebrain cholinergic system parallels cognitive impairment in patients at high risk of developing Alzheimer's disease. Cereb Cortex 2010;20:1685–1695.

191 Pihlajamaki M, Jauhiainen AM, Soininen H: Structural and functional MRI in mild cognitive impairment. Curr Alzheimer Res 2009;6:179–185.

192 Petrella JR, Sheldon FC, Prince SE, Calhoun VD, Doraiswamy PM: Default mode network connectivity in stable vs. progressive mild cognitive impairment. Neurology 2011;76:511–517.

193 Sperling RA, Dickerson BC, Pihlajamaki M, Vannini P, LaViolette PS, Vitolo OV, et al: Functional alterations in memory networks in early Alzheimer's disease. Neuromolecular Med 2010;12:27–43.

194 Filippini N, Rao A, Wetten S, Gibson RA, Borrie M, Guzman D, et al: Anatomically-distinct genetic associations of APOE ε4 allele load with regional cortical atrophy in Alzheimer's disease. Neuroimage 2009;44:724–728.

195 Sporns O, Honey CJ: Small worlds inside big brains. Proc Natl Acad Sci USA 2006;103:19219–19220.

196 Skudlarski P, Jagannathan K, Calhoun VD, Hampson M, Skudlarska BA, Pearlson G: Measuring brain connectivity: diffusion tensor imaging validates resting state temporal correlations. Neuroimage 2008;43:554–561.
197 Skudlarski P, Skudlarska BA, Pearlson G: in Ann Meet Am Geriatr Soc, Chicago, IL, 2009, pp S34–S35.
198 Park DC, Reuter-Lorenz P: The adaptive brain: aging and neurocognitive scaffolding. Annu Rev Psychol 2009;60:173–196.
199 Nordberg A, Rinne JO, Kadir A, Langstrom B: The use of PET in Alzheimer disease. Nat Rev Neurol 2010;6:78–87.
200 Herholz K: Molecular imaging in Alzheimer's disease. Eur Neurol Rev 2011;6:16–20.
201 Klunk WE, Engler H, Nordberg A, Wang Y, Blomqvist G, Holt DP, et al: Imaging brain amyloid in Alzheimer's disease with Pittsburgh Compound B. Ann Neurol 2004;55:306–319.
202 Edison P, Archer HA, Hinz R, Hammers A, Pavese N, Tai YF, et al: Amyloid, hypometabolism, and cognition in Alzheimer disease: an [^{11}C]PIB and [^{18}F]FDG-PET study. Neurology 2007;68:501–508.
203 Kemppainen NM, Aalto S, Wilson IA, Nagren K, Helin S, Bruck A, et al: Voxel-based analysis of PET amyloid ligand [^{11}C]PIB uptake in Alzheimer disease. Neurology 2006;67:1575–1580.
204 Nordberg A: PET imaging of amyloid in Alzheimer's disease. Lancet Neurol 2004;3:519–527.
205 Devanand DP, Mikhno A, Pelton GH, Cuasay K, Pradhaban G, Dileep Kumar JS, et al: Pittsburgh Compound B (^{11}C-PIB) and fluorodeoxyglucose (^{18}F-FDG)-PET in patients with Alzheimer disease, mild cognitive impairment, and healthy controls. J Geriatr Psychiatry Neurol 2010;23:185–198.
206 Cairns NJ, Ikonomovic MD, Benzinger T, Storandt M, Fagan AM, Shah AR, et al: Absence of Pittsburgh Compound B detection of cerebral amyloid-β in a patient with clinical, cognitive, and cerebrospinal fluid markers of Alzheimer disease: a case report. Arch Neurol 2009;66:1557–1562.
207 Rowe CC, Ellis KA, Rimajova M, Bourgeat P, Pike KE, Jones G, et al: Amyloid imaging results from the Australian Imaging, Biomarkers and Lifestyle (AIBL) study of aging. Neurobiol Aging 2010;31:1275–1283.
208 Mormino EC, Kluth JT, Madison CM, Rabinovici GD, Baker SL, Miller BL, et al: Episodic memory loss is related to hippocampal-mediated β-amyloid deposition in elderly subjects. Brain 2009;132:1310–1323.
209 Pike KE, Savage G, Villemagne VL, Ng S, Moss SA, Maruff P, et al: Beta-amyloid imaging and memory in non-demented individuals: evidence for preclinical Alzheimer's disease. Brain 2007;130:2837–2844.
210 Morris JC, Roe CM, Grant EA, Head D, Storandt M, Goate AM, et al: Pittsburgh compound B imaging and prediction of progression from cognitive normality to symptomatic Alzheimer disease. Arch Neurol 2009;66:1469–1475.
211 Reiman EM, Chen K, Liu X, Bandy D, Yu M, Lee W, et al: Fibrillar amyloid-β burden in cognitively normal people at three levels of genetic risk for Alzheimer's disease. Proc Natl Acad Sci USA 2009; 106:6820–6825.
212 Forsberg A, Engler H, Almkvist O, Blomquist G, Hagman G, Wall A, et al: PET imaging of amyloid deposition in patients with mild cognitive impairment. Neurobiol Aging 2008;29:1456–1465.
213 Okello A, Koivunen J, Edison P, Archer HA, Turkheimer FE, Nagren K, et al: Conversion of amyloid positive and negative MCI to AD over 3 years: an ^{11}C-PIB PET study. Neurology 2009;73: 754–760.
214 Jack CR Jr, Wiste HJ, Vemuri P, Weigand SD, Senjem ML, Zeng G, et al: Brain β-amyloid measures and magnetic resonance imaging atrophy both predict time-to-progression from mild cognitive impairment to Alzheimer's disease. Brain 2010;133:3336–3348.
215 Lowe VJ, Kemp BJ, Jack CR Jr, Senjem M, Weigand S, Shiung M, et al: Comparison of ^{18}F-FDG and PiB PET in cognitive impairment. J Nucl Med 2009;50: 878–886.
216 Cohen AD, Price JC, Weissfeld LA, James J, Rosario BL, Bi W, et al: Basal cerebral metabolism may modulate the cognitive effects of Aβ in mild cognitive impairment: an example of brain reserve. J Neurosci 2009;29:14770–14778.
217 Jack CR Jr, Lowe VJ, Weigand SD, Wiste HJ, Senjem ML, Knopman DS, et al: Serial PIB and MRI in normal, mild cognitive impairment and Alzheimer's disease: implications for sequence of pathological events in Alzheimer's disease. Brain 2009;132: 1355–1365.
218 Rabinovici GD, Furst AJ, O'Neil JP, Racine CA, Mormino EC, Baker SL, et al: ^{11}C-PIB PET imaging in Alzheimer disease and frontotemporal lobar degeneration. Neurology 2007;68:1205–1212.
219 Gomperts SN, Rentz DM, Moran E, Becker JA, Locascio JJ, Klunk WE, et al: Imaging amyloid deposition in Lewy body diseases. Neurology 2008; 71:903–910.
220 Edison P, Rowe CC, Rinne JO, Ng S, Ahmed I, Kemppainen N, et al: Amyloid load in Parkinson's disease dementia and Lewy body dementia measured with [^{11}C]PIB positron emission tomography. J Neurol Neurosurg Psychiatry 2008;79:1331–1338.

221 Ly JV, Donnan GA, Villemagne VL, Zavala JA, Ma H, O'Keefe G, et al: ^{11}C-PIB binding is increased in patients with cerebral amyloid angiopathy-related hemorrhage. Neurology 2010;74:487–493.

222 Sabri O, Gertz H, Dresel S, Heuser I, Bartenstein P, Buerger K, et al: Multicentre phase 2 trial on florbetaben for β-amyloid brain PET in Alzheimer disease. J Nucl Med 2010;51(suppl):384.

223 Wong DF, Rosenberg PB, Zhou Y, Kumar A, Raymont V, Ravert HT, et al: In vivo imaging of amyloid deposition in Alzheimer disease using the radioligand ^{18}F-AV-45 (flobetapir F18). J Nucl Med 2010;51:913–920.

224 Rowe CC, Ackerman U, Browne W, Mulligan R, Pike KL, O'Keefe G, et al: Imaging of amyloid-β in Alzheimer's disease with ^{18}F-BAY-94-9172, a novel PET tracer: proof of mechanism. Lancet Neurol 2008;7:129–135.

225 Vandenberghe R, Van Laere K, Ivanoiu A, Salmon E, Bastin C, Triau E, et al: ^{18}F-flutemetamol amyloid imaging in Alzheimer disease and mild cognitive impairment: a phase 2 trial. Ann Neurol 2010;68: 319–329.

226 Clark CM, Schneider JA, Bedell BJ, Beach TG, Bilker WB, Mintun MA, et al: Use of florbetapir-PET for imaging β-amyloid pathology. JAMA 2011; 305:275–283.

227 Sokoloff L: Relation between physiological function and energy metabolism in the central nervous system. J Neurochem 1977;29:13–26.

228 Kasischke KA, Vishwasrao HD, Fisher PJ, Zipfel WR, Webb WW: Neural activity triggers neuronal oxidative metabolism followed by astrocytic glycolysis. Science 2004;305:99–103.

229 Herholz K: Cerebral glucose metabolism in preclinical and prodromal Alzheimer's disease. Expert Rev Neurother 2010;10:1667–1673.

230 Landau SM, Harvey D, Madison CM, Koeppe RA, Reiman EM, Foster NL, et al: Associations between cognitive, functional, and FDG-PET measures of decline in AD and MCI. Neurobiol Aging 2011;32: 1207–1218.

231 Minoshima S, Frey KA, Koeppe RA, Foster NL, Kuhl DE: A diagnostic approach in Alzheimer's disease using three-dimensional stereotactic surface projections of fluorine-18-FDG-PET. J Nucl Med 1995;36:1238–1248.

232 Herholz K, Salmon E, Perani D, Baron JC, Holthoff V, Frolich L, et al: Discrimination between Alzheimer dementia and controls by automated analysis of multicenter FDG-PET. Neuroimage 2002;17:302–316.

233 Haense C, Herholz K, Heiss WD: Validation of an automated FDG-PET analysis to discriminate patients with Alzheimer's disease from normal subjects. J Nucl Med 2008;49(suppl 1):34P.

234 Mosconi L, Tsui WH, Herholz K, Pupi A, Drzezga A, Lucignani G, et al: Multicenter standardized ^{18}F-FDG-PET diagnosis of mild cognitive impairment, Alzheimer's disease, and other dementias. J Nucl Med 2008;49:390–398.

235 Winkeler A, Boisgard R, Martin A, Tavitian B: Radioisotopic imaging of neuroinflammation. J Nucl Med 2010;51:1–4.

236 Mattiace LA, Davies P, Yen SH, Dickson DW: Microglia in cerebellar plaques in Alzheimer's disease. Acta Neuropathol 1990;80:493–498.

237 Perry VH, Nicoll JA, Holmes C: Microglia in neurodegenerative disease. Nat Rev Neurol 2010;6: 193–201.

238 Cagnin A, Brooks DJ, Kennedy AM, Gunn RN, Myers R, Turkheimer FE, et al: In-vivo measurement of activated microglia in dementia. Lancet 2001;358:461–467.

239 Banati RB, Newcombe J, Gunn RN, Cagnin A, Turkheimer F, Heppner F, et al: The peripheral benzodiazepine binding site in the brain in multiple sclerosis: quantitative in vivo imaging of microglia as a measure of disease activity. Brain 2000;123:2321–2337.

240 Groom GN, Junck L, Foster NL, Frey KA, Kuhl DE: PET of peripheral benzodiazepine binding sites in the microgliosis of Alzheimer's disease. J Nucl Med 1995;36:2207–2210.

241 Edison P, Archer HA, Gerhard A, Hinz R, Pavese N, Turkheimer FE, et al: Microglia, amyloid, and cognition in Alzheimer's disease: an [^{11}C](R)PK11195-PET and [^{11}C]PIB-PET study. Neurobiol Dis 2008; 32:412–419.

242 Chauveau F, Boutin H, Van Camp N, Dolle F, Tavitian B: Nuclear imaging of neuroinflammation: a comprehensive review of [^{11}C]PK11195 challengers. Eur J Nucl Med Mol Imaging 2008;35:2304–2319.

243 Perry EK, Haroutunian V, Davis KL, Levy R, Lantos P, Eagger S, et al: Neocortical cholinergic activities differentiate Lewy body dementia from classical Alzheimer's disease. Neuroreport 1994;5:747–749.

244 Namba H, Irie T, Fukushi K, Iyo M: In vivo measurement of acetylcholinesterase activity in the brain with a radioactive acetylcholine analog. Brain Res 1994;667:278–282.

245 Kilbourn MR, Snyder SE, Sherman PS, Kuhl DE: In vivo studies of acetylcholinesterase activity using a labeled substrate, N-[^{11}C]methylpiperdin-4-yl propionate ([^{11}C]PMP). Synapse 1996;22:123–131.

246 Herholz K: Acetylcholine esterase activity in mild cognitive impairment and Alzheimer's disease. Eur J Nucl Med Mol Imaging 2008;35(suppl 1):S25–S29.

247 Mesulam M: Neuroanatomy of cholinesterases in the normal human brain and in Alzheimer's disease; in Giacobini E (ed): Cholinesterases and Cholinesterase Inhibitors. London, Dunitz, 2000, pp 121–137.
248 Haense C, Herholz K, Jagust WJ, Heiss WD: Performance of FDG-PET for detection of Alzheimer's disease in two independent multicenter samples (NEST-DD and ADNI). Dement Geriatr Cogn Disord 2009;28:259–266.
249 Nordberg A, Lundqvist H, Hartvig P, Lilja A, Langstrom B: Kinetic analysis of regional (S) (–)^{11}C-nicotine binding in normal and Alzheimer brains – in vivo assessment using positron emission tomography. Alzheimer Dis Assoc Disord 1995;9: 21–27.
250 Kadir A, Almkvist O, Wall A, Langstrom B, Nordberg A: PET imaging of cortical ^{11}C-nicotine binding correlates with the cognitive function of attention in Alzheimer's disease. Psychopharmacology (Berl) 2006;188:509–520.
251 Bottlaender M, Valette H, Roumenov D, Dolle F, Coulon C, Ottaviani M, et al: Biodistribution and radiation dosimetry of ^{18}F-fluoro-A-85380 in healthy volunteers. J Nucl Med 2003;44:596–601.
252 Mamede M, Ishizu K, Ueda M, Mukai T, Iida Y, Fukuyama H, et al: Quantification of human nicotinic acetylcholine receptors with ^{123}I-5IA SPECT. J Nucl Med 2004;45:1458–1470.
253 Pimlott SL, Piggott M, Owens J, Greally E, Court JA, Jaros E, et al: Nicotinic acetylcholine receptor distribution in Alzheimer's disease, dementia with Lewy bodies, Parkinson's disease, and vascular dementia: in vitro binding study using 5-[^{125}I]-A-85380. Neuropsychopharmacology 2004;29:108–116.
254 Terriere E, Dempsey MF, Herrmann LL, Tierney KM, Lonie JA, O'Carroll RE, et al: 5-^{123}I-A-85380 binding to the $\alpha_4\beta_2$-nicotinic receptor in mild cognitive impairment. Neurobiol Aging 2010;31: 1885–1893.
255 Meyer PM, Strecker K, Kendziorra K, Becker G, Hesse S, Woelpl D, et al: Reduced $\alpha_4\beta_2$*-nicotinic acetylcholine receptor binding and its relationship to mild cognitive and depressive symptoms in Parkinson disease. Arch Gen Psychiatry 2009;66: 866–877.
256 Horti AG, Gao Y, Kuwabara H, Dannals RF: Development of radioligands with optimized imaging properties for quantification of nicotinic acetylcholine receptors by positron emission tomography. Life Sci 2010;86:575–584.
257 Brooks DJ: Advances in imaging Parkinson's disease. Curr Opin Neurol 1997;10:327–331.
258 Hu XS, Okamura N, Arai H, Higuchi M, Matsui T, Tashiro M, et al: ^{18}F-fluorodopa PET study of striatal dopamine uptake in the diagnosis of dementia with Lewy bodies. Neurology 2000;55:1575–1577.
259 Hilker R, Thomas A, Klein JC, Weisenbach S, Kalbe E, Burghaus L, et al: Dementia in Parkinson's disease: functional imaging of cholinergic and dopaminergic pathways. Neurology 2005;65:1716–1722.
260 de Leon MJ, Golomb J, George AE, Convit A, Tarshish CY, McRae T, et al: The radiologic prediction of Alzheimer disease: the atrophic hippocampal formation. AJNR Am J Neuroradiol 1993;14: 897–906.
261 Scheltens P, Leys D, Barkhof F, Huglo D, Weinstein HC, Vermersch P, et al: Atrophy of medial temporal lobes on MRI in 'probable' Alzheimer's disease and normal ageing: diagnostic value and neuropsychological correlates. J Neurol Neurosurg Psychiatry 1992;55:967–972.
262 Jack CR Jr, Petersen RC, O'Brien PC, Tangalos EG: MR-based hippocampal volumetry in the diagnosis of Alzheimer's disease. Neurology 1992;42:183–188.
263 Fischl B, Salat DH, Busa E, Albert M, Dieterich M, Haselgrove C, et al: Whole brain segmentation: automated labeling of neuroanatomical structures in the human brain. Neuron 2002;33:341–355.
264 Dale AM, Fischl B, Sereno MI: Cortical surface-based analysis. I. Segmentation and surface reconstruction. Neuroimage 1999;9:179–194.
265 Barnes J, Foster J, Boyes RG, Pepple T, Moore EK, Schott JM, et al: A comparison of methods for the automated calculation of volumes and atrophy rates in the hippocampus. Neuroimage 2008;40:1655–1671.
266 Morra JH, Tu Z, Apostolova LG, Green AE, Avedissian C, Madsen SK, et al: Automated 3D mapping of hippocampal atrophy and its clinical correlates in 400 subjects with Alzheimer's disease, mild cognitive impairment, and elderly controls. Hum Brain Mapp 2009;30:2766–2788.
267 Kim H, Besson P, Colliot O, Bernasconi A, Bernasconi N: Surface-based vector analysis using heat equation interpolation: a new approach to quantify local hippocampal volume changes. Med Image Comput Comput Assist Interv 2008;11: 1008–1015.
268 Mouiha A, Duchesne S: Hippocampal atrophy rates in Alzheimer's disease: automated segmentation variability analysis. Neurosci Lett (E-pub ahead of print).
269 Wolz R, Heckemann RA, Aljabar P, Hajnal JV, Hammers A, Lotjonen J, et al: Measurement of hippocampal atrophy using 4D graph-cut segmentation: application to ADNI. Neuroimage 2010;52: 109–118.

270 Leung KK, Barnes J, Ridgway GR, Bartlett JW, Clarkson MJ, Macdonald K, et al: Automated cross-sectional and longitudinal hippocampal volume measurement in mild cognitive impairment and Alzheimer's disease. Neuroimage 2010;51:1345–1359.

271 Chupin M, Hammers A, Liu RS, Colliot O, Burdett J, Bardinet E, et al: Automatic segmentation of the hippocampus and the amygdala driven by hybrid constraints: method and validation. Neuroimage 2009;46:749–761.

272 Vemuri P, Gunter JL, Senjem ML, Whitwell JL, Kantarci K, Knopman DS, et al: Alzheimer's disease diagnosis in individual subjects using structural MR images: validation studies. Neuroimage 2008;39:1186–1197.

273 Davatzikos C, Fan Y, Wu X, Shen D, Resnick SM: Detection of prodromal Alzheimer's disease via pattern classification of magnetic resonance imaging. Neurobiol Aging 2008;29:514–523.

274 Stonnington CM, Tan G, Kloppel S, Chu C, Draganski B, Jack CR Jr, et al: Interpreting scan data acquired from multiple scanners: a study with Alzheimer's disease. Neuroimage 2008;39:1180–1185.

275 Teipel SJ, Bayer W, Alexander GE, Bokde AL, Zebuhr Y, Teichberg D, et al: Regional pattern of hippocampus and corpus callosum atrophy in Alzheimer's disease in relation to dementia severity: evidence for early neocortical degeneration. Neurobiol Aging 2003;24:85–94.

276 Frisoni GB, Fox NC, Jack CR Jr, Scheltens P, Thompson PM: The clinical use of structural MRI in Alzheimer disease. Nat Rev Neurol 2010;6:67–77.

277 Jack CR Jr, Bernstein MA, Fox NC, Thompson P, Alexander G, Harvey D, et al: The Alzheimer's Disease Neuroimaging Initiative (ADNI): MRI methods. J Magn Reson Imaging 2008;27:685–691.

278 Haacke EM, Xu Y, Cheng YC, Reichenbach JR: Susceptibility-weighted imaging. Magn Reson Med 2004;52:612–618.

279 Jack CR Jr, Bernstein MA, Borowski BJ, Gunter JL, Fox NC, Thompson PM, et al: Update on the magnetic resonance imaging core of the Alzheimer's disease neuroimaging initiative. Alzheimers Dement 2010;6:212–220.

280 Hajnal JV, Hill DLG, Hawkes DJ: Medical Image Registration. New York, CRC Press, 2001.

281 Han X, Jovicich J, Salat D, van der Kouwe A, Quinn B, Czanner S, et al: Reliability of MRI-derived measurements of human cerebral cortical thickness: the effects of field strength, scanner upgrade and manufacturer. Neuroimage 2006;32:180–194.

282 Narayana PA, Brey WW, Kulkarni MV, Sievenpiper CL: Compensation for surface coil sensitivity variation in magnetic resonance imaging. Magn Reson Imaging 1988;6:271–274.

283 Clarkson MJ, Ourselin S, Nielsen C, Leung KK, Barnes J, Whitwell JL, et al: Comparison of phantom and registration scaling corrections using the ADNI cohort. Neuroimage 2009;47:1506–1513.

284 Gunter JL, Bernstein MA, Borowski BJ, Ward CP, Britson PJ, Felmlee JP, et al: Measurement of MRI scanner performance with the ADNI phantom. Med Phys 2009;36:2193–2205.

285 Leow AD, Klunder AD, Jack CR Jr, Toga AW, Dale AM, Bernstein MA, et al: Longitudinal stability of MRI for mapping brain change using tensor-based morphometry. Neuroimage 2006;31:627–640.

286 Boyes RG, Gunter JL, Frost C, Janke AL, Yeatman T, Hill DL, et al: Intensity non-uniformity correction using N3 on 3-T scanners with multichannel phased array coils. Neuroimage 2008;39:1752–1762.

287 Teipel SJ, Ewers M, Wolf S, Jessen F, Kolsch H, Arlt S, et al: Multicentre variability of MRI-based medial temporal lobe volumetry in Alzheimer's disease. Psychiatry Res 2010;182:244–250.

288 Cover KS, van Schijndel RA, van Dijk BW, Redolfi A, Knol DL, Frisoni GB, et al: Assessing the reproducibility of the SienaX and Siena brain atrophy measures using the ADNI back-to-back MP-RAGE MRI scans. Psychiatry Res 2011;193:182–190.

289 Frisoni GB, Jack CR: Harmonization of magnetic resonance-based manual hippocampal segmentation: a mandatory step for wide clinical use. Alzheimers Dement 2011;7:171–174.

290 McMahon PM, Araki SS, Sandberg EA, Neumann PJ, Gazelle GS: Cost-effectiveness of PET in the diagnosis of Alzheimer disease. Radiology 2003;228:515–522.

291 Meindl T, Teipel S, Elmouden R, Mueller S, Koch W, Dietrich O, et al: Test-retest reproducibility of the default-mode network in healthy individuals. Hum Brain Mapp 2009.

Prof. Stefan Teipel
Department of Psychiatry, University Rostock and DZNE
German Center for Neurodegenerative Diseases
Gehlsheimer Strasse 20, DE–18147 Rostock (Germany)
Tel. +49 381 494 9471, E-Mail stefan.teipel@med.uni-rostock.de

Hampel H, Carrillo MC (eds): Alzheimer's Disease – Modernizing Concept, Biological Diagnosis and Therapy.
Adv Biol Psychiatry. Basel, Karger, 2012, vol 28, pp 115–121

Changing Diagnostic Concepts of Alzheimer's Disease

Guy M. McKhann[a] · Marilyn S. Albert[a] · Reisa A. Sperling[b]

[a]Department of Neurology, Johns Hopkins School of Medicine, Baltimore, Md., and
[b]Department of Neurology, Brigham & Women's Hospital, Harvard Medical School, Boston, Mass., USA

Abstract

Diagnostic concepts of Alzheimer's disease have evolved over the past few decades. There is now a consensus that Alzheimer's disease pathophysiological processes develop slowly over many years and, thus, it is appropriate for diagnostic approaches to consider the continuum of disease, from presymptomatic individuals, to those with mild cognitive impairment, to patients with dementia. Several diagnostic criteria have been proposed using this perspective. This chapter focuses on the National Institute on Aging – Alzheimer's Association criteria published in 2011 and compares them with two other diagnostic criteria – those proposed by the International Working Group and those proposed by the DSM-5 workgroup.

In 1984 a working group outlined research criteria for the diagnosis of Alzheimer's disease (AD) [1]. These proposed criteria, which were intended to be provisional, lasted for over 26 years, much to the surprise of the initial participants. However, by 2009 it became apparent to a group convened by the National Institute on Aging (NIA) and the Alzheimer's Association (AA) that the original criteria should be re-evaluated in terms of how our concepts of AD have changed in the intervening years.

The new criteria that ultimately emerged from this process are based on several concepts about AD that did not exist in 1984.

(1) The AD process is a continuum from presymptomatic individuals, to those with mild cognitive impairment, to patients with dementia. It has become apparent that the pathological hallmarks of the disease (i.e. amyloid plaques and neurofibrillary tangles) are present years before cognitive symptoms are recognized. The discussion of AD in 1984 emphasized AD as a dementia; the revised criteria emphasize this continuum.

(2) Impairment in memory is not always the primary cognitive deficit in patients with AD. While most patients (perhaps more than 80%) present with episodic memory deficits (i.e. difficulty learning and retaining new information), other patients present

with problems with language, spatial ability or executive function (i.e. impaired reasoning, planning, problem-solving).

(3) Other causes of cognitive decline have been characterized that were barely recognized in 1984, including dementia with Lewy bodies and frontotemporal lobar degeneration. Additionally, causes of cognitive decline that were widely acknowledged in 1984 commonly coexist with AD pathology, particularly in older individuals (e.g. vascular disease).

(4) Biomarkers, which did not exist in 1984, have been developed which aid in diagnosis, and facilitate the prediction of likelihood of progression to greater degrees of cognitive impairment.

In light of these changes, in 2009, the NIA and AA established three working groups to develop new criteria for AD. Each of the groups was asked to focus on one of the three phases of AD: (1) the presymptomatic or 'preclinical' phase, (2) the symptomatic, predementia phase, and (3) the dementia phase. Each of the working groups were international in composition, and included individuals with a range of expertise (i.e. neurologists, psychiatrists, neuropsychologists, radiologists, biostatisticians) from both academia and the pharmaceutical industry.

The working groups that dealt with the symptomatic phase of disease were specifically asked to include criteria that would have wide application in a range of settings, including primary care physicians in countries around the world, specialists in tertiary care centers, clinical researchers in academic settings, and those planning clinical trials.

Clinical Criteria for the Dementia of Alzheimer's Disease

The clinical criteria for AD dementia in the newly published criteria are very similar to those outlined in 1984. The Core Clinical Criteria for the dementia of AD in the new criteria [2] include:

(1) An insidious onset of months to years, not hours or days.

(2) Clear-cut history of progression of cognitive decline. This history is usually obtained from an informant. Cognitive decline can be affirmed by neuropsychological testing.

(3) The most common presentation is the amnestic form, involving an impairment in episodic memory, i.e. the ability to learn and retain new information. However, some patients have initial involvement of other cognitive domains, such as language, or executive functioning (impaired reasoning, planning and problem solving).

(4) Differential diagnosis of AD dementia. Dementia due to AD is less likely when there are symptoms of other potential causes of dementia, such as vascular disease or frontotemporal dementia. In addition, other factors, such as depression, must be considered.

Clinical Criteria for Mild Cognitive Impairment due to Alzheimer's Disease

The term mild cognitive impairment (MCI) represents those who have some cognitive impairment, but have not progressed to dementia. As the charge to the working group concerned AD, the focus was, by definition, on 'MCI due to AD'. The Core Clinical Criteria in the new criteria [3] include:

(1) Concern regarding a change in cognition. This concern may be identified by the patient, an informant or the clinician.

(2) Impairment in one or more cognitive domains. This impairment is greater than one would expect for the patient's age or educational background.

(3) Preservation of independence in functional activities. There may be some mild problems with more complex activities, but those with MCI maintain their independence in daily life, requiring only minimal assistance.

(4) Not demented. The degree of decline in cognition and function is not consistent with dementia. There are no exact transition points that define when an individual has progressed from the MCI phase to the dementia phase. It is a question of clinical judgment.

(5) Differential diagnosis of MCI due to AD. As with dementia, there are causes of MCI other than AD pathology. It is necessary to rule out other systemic and brain diseases that can alter cognitive function, such as vascular disease, other neurodegenerative diseases and depression.

Research Criteria for Preclinical Alzheimer's Disease

The preclinical phase of the disease is primarily defined by changes in biomarkers (as discussed below), but may be accompanied by evidence of change from one's own baseline cognitive status or poor performance on challenging cognitive tests. The term 'preclinical' was selected, as opposed to 'presymptomatic', to recognize that individuals may demonstrate very subtle changes in cognition and behavior that may precede MCI by several years, but would not typically be brought to clinical attention. The recommendations for the study of preclinical AD [4] are intended purely for research purposes and do not have any clinical utility at the present time.

Biomarkers

The most novel aspect of the three sets of recommendations is the application of biomarkers. Biomarkers are measures, usually of a biological fluid or imaging, that systematically relate to the presence and progression of the underlying disease process. As outlined in the work group reports, biomarkers for AD fall into two major categories:

(1) Biomarkers that reflect the accumulation of the amyloid-β peptide (Aβ). Aβ is a component of the plagues that are diagnostic of AD on autopsy. There are currently two biomarkers of Aβ: the detection of low levels of Aβ in cerebrospinal fluid (CSF) [5] and the abnormal accumulation of an amyloid imaging agent [6] on positron emission tomography (PET) scans.

(2) Biomarkers of neuronal injury. There are two classes of such biomarkers. Those related to the detection of tau, which is a component of the tangles that are diagnostic of AD on autopsy, and imaging measures reflecting neuronal injury. Elevated CSF tau (both total tau and phosphorylated tau) is typically seen in AD [5] (but also occurs in several other neurologic conditions, hence is less specific for AD than is Aβ). Imaging measures that reflect neuronal injury include: decreased fluorodeoxyglucose uptake on PET scans in a specific topographic pattern [6] involving the temporoparietal cortex; atrophy on structural magnetic resonance imaging, again in a specific topographic pattern [7] typically involving the medial, basal and lateral temporal lobe, and/or medial parietal cortex.

For the diagnostic criteria pertaining to the symptomatic phases of disease, a set of algorithms were proposed by which biomarkers from the two categories mentioned above could be examined individually, or in combination, for two overarching purposes:

(1) To enhance the certainty that the underlying disease is AD.

(2) To estimate the likelihood of progression to the next stage of the disease (e.g. from MCI to dementia).

It was proposed that when biomarkers from both categories are abnormal (i.e. biomarkers reflecting Aβ and biomarkers of neuronal injury were both positive), then the individual had the highest likelihood of having AD pathology as the underlying cause of their symptoms and, thus, had the highest likelihood of progression over time. When biomarkers from both categories were normal, then the individual had the lowest likelihood of having AD pathology as the underlying cause of their symptoms and, therefore, had the lowest likelihood of progression over time. Intermediate levels of likelihood were proposed when biomarkers from one of the two categories were positive.

For individuals in the preclinical phase of AD, the working group proposed a three-stage model of likelihood of progression towards MCI and AD dementia. This model is based on the hypothesis that progression of AD biomarkers follow an ordered temporal pattern [8, 9] which precedes clinical symptomatology. The preclinical working group report summarized evidence suggesting that amyloid biomarkers become abnormal approximately 10 years prior to the first clinical symptoms, and that biomarkers of neuronal injury become dynamic at a later point. They also suggested that biomarkers of neuronal injury may begin to change shortly before clinical symptoms first appear, and that subtle cognitive or behavioral symptomatology may be present for several years prior to the time that 'MCI due to AD' can be easily recognized. The working group emphasized that the proposed framework does not prove that Aβ is the initiating factor for the disease, and that Aβ may be 'necessary but not sufficient' to result in the clinical phases of the illness. The evidence to date does, however, suggest that these different categories of biomarkers may be useful for tracking progression in the earliest stages of AD.

It is largely because the framework for biomarkers proposed by the working groups was based on data that is continuing to emerge in the field, that all of the reports emphasized the need for testing the hypothetical models proposed. For example, there is limited information about likelihood of progression when biomarker findings are discordant. Additionally, few studies have examined biomarkers in community settings. Further, there is not yet a standardization of techniques and levels between laboratories, such as for CSF Aβ or tau. For all of these reasons, the use of biomarkers was recommended primarily for research purposes, and not for clinical practice.

It was envisioned that one of the research settings in which this framework for biomarkers would be tested was in the design of clinical trials. This is particularly important in the context of early intervention since individuals who meet clinical criteria for MCI can be quite heterogeneous. Thus, biomarkers could be used to increase the likelihood that subjects recruited into a clinical trial had AD as the underlying cause of their symptoms. For example, if the drug to be tested was directly aimed at altering levels of Aβ, and one had established that the subjects enrolled in the trial had evidence of amyloid accumulation (either in CSF or on a PET scan), then one could be more certain that the clinical trial was testing the pathological process it was aimed at. These biomarkers may also prove useful in defining preclinical AD populations for secondary prevention trials aimed at slowing the progression of neuronal injury and the emergence of clinical symptoms.

Comparison of NIA-AA Criteria with Other Diagnostic Frameworks

Two other diagnostic frameworks have been, or are about to be, proposed for individuals with AD. One of these [10, 11] was designed to reflect the evolution in the concepts of AD described above. A second will be incorporated into the general reevaluation of diagnostic criteria for a broad range of psychiatric and neurological disorders sponsored by the American Psychiatric Association (i.e. the DSM-5).

The criteria proposed by Dubois and colleagues [10, 11] are conceptually similar in many respects to the NIA-AA criteria. The Dubois criteria, like the NIA-AA criteria, emphasize that the AD process is a continuum from the preclinical to the prodromal to the dementia phases of AD. They, likewise, emphasize the importance of biomarkers that can be used to increase the likelihood that the cognitive syndrome observed is due to AD pathology in the brain.

Additionally, the Core Clinical Criteria in both proposals emphasize that the majority of individuals with AD have an episodic memory impairment as the most common presenting symptom. However, the Dubois criteria focus heavily on this 'amnestic' presentation of AD, concluding that the 'atypical' presentations of AD should not be considered to meet criteria for either prodromal AD or AD dementia. This contrasts with the NIA-AA criteria for both 'MCI due to AD' and 'AD dementia', which allow individuals with atypical presentations to meet their Core Clinical Criteria. This difference in approach is due, in large part, to the fact that the NIA-AA workgroups focused on criteria for symptomatic individuals were intent on developing Core Clinical Criteria that could be applied in a broad range of clinical and community settings, whereas the Dubois criteria are primarily focused on identifying individuals with a strong likelihood of AD pathology who would be eligible for research studies, particularly clinical trials. The application of biomarkers also differs between the two sets of criteria. The NIA-AA criteria for 'AD dementia' and 'MCI due to AD' outline various levels of certainty for the diagnosis, depending upon which biomarkers are found to be positive. The Dubois criteria do not use biomarkers of varying types to differentiate between levels of certainty; thus any individual with an 'amnestic' presentation and a positive biomarker fulfills the criteria for AD. Lastly, the Dubois criteria emphasize the importance of establishing the presence of an episodic memory deficit by using a memory test that includes selective cueing to establish that each item has been registered in memory, while the NIA-AA criteria describe a wide range of memory tests that can be employed in assessing patients.

Additionally, there are differences in terminology between the two sets of criteria that have substantial practical consequences. For example, the NIA-AA criteria retain a two-step process by which an individual is first examined to determine if they meet the Core Clinical Criteria and then biomarkers are used to increase the likelihood that AD pathology is the underlying cause of the cognitive impairment, whereas this is not the case for the Dubois criteria. For example, the term MCI in the lexicon of Dubois et al. [10] specifically excludes individuals who would meet the NIA-AA criteria for 'MCI due to AD'. The terminology for asymptomatic individuals with AD pathology also differs. The NIA-AA workgroup [4] uses the term 'preclinical AD' for this phase of disease, while the Dubois lexicon has two categories within the preclinical state: 'asymptomatic at-risk state for AD' and 'presymptomatic AD', the latter category reserved for individuals with autosomal dominant genetic mutations.

The DSM-5 workgroup will recommend yet another set of terms (www.dsm5.org), to take the place of the criteria for 'Delirium, Dementia, Amnestic and Other Geriatric Cognitive Disorders' contained in DSM-IV. Individuals with dementia will fit within the category of 'major neurocognitive disorder' and individuals with MCI will fit within the category of 'mild neurocognitive disorder'. The description of the clinical syndromes for both MCI and dementia are, however, fairly similar in the DSM-5 and NIA-AA criteria. The approach to cognitive testing that is recommended is also similar in that a range of cognitive tests are described, but no one particular test, or set of tests, is recommended.

Conclusion

The primary goal of the NIA-AA recommendations was to emphasize that the development of AD takes many years, and that individuals with evidence of pathology have a range of symptoms. A secondary goal was to realign the clinical and research diagnostic approach with potential therapies. The members of the three working groups were concerned that if one waits for the appearance of dementia to introduce treatment, it may be too late, since neuronal damage is extensive by this time. It would be as if we tried to prevent heart disease in only those who have had a myocardial infarction. It is hoped that establishing criteria for the earlier phases of the disease will pave the way for earlier intervention, and the successful development of disease modifying therapies.

References

1 McKhann G, Drachman D, Folstein M, et al: Clinical diagnosis of Alzheimer's disease: report of the NINCDS-ADRDA Work Group under the auspices of Department of Health and Human Services Task Force on Alzheimer's Disease. Neurol 1984;34:939–944.

2 McKhann G, Knopman D, Chertkow H, Hyman B, Jack C, Kawas C, et al: The National Institute on Aging – Alzheimer's Association Workgroup on the diagnosis of dementia due to Alzheimer's disease. Alzheimers Dement 2011;7:232–239.

3 Albert MS, DeKosky S, Dickson D, Dubois B, Feldman HH, Fox NC, et al: The diagnosis of mild cognitive impairment due to Alzheimer's disease: recommendations from the National Institute on Aging-Alzheimer's Association workgroups on diagnostic guidelines for Alzheimer's disease. Alzheimers Dement 2011;7: 270–279.

4 Sperling R, Aisen P, Beckett L, Bennett D, Craft S, Fagan A, et al: Toward defining the preclinical stages of Alzheimer's disease: National Institute on Aging and Alzheimer's Association Workgroup. Alzheimers Dement 2011;7:280–292.

5 Shaw L, Vanderstichele H, Knapik-Czajka M, Clark C, Aisen P, Petersen R, Blennow K, Soares H, Simon A, Lewczuk P, Dean R, Siemers E, Potter W, Lee V, Trojanowski J: Cerebrospinal fluid biomarker signature in Alzheimer's disease neuroimaging initiative subjects. Ann Neurol 2009;65:403–413.

6 Jagust W: Positron emission tomography and magnetic resonance imaging in the diagnosis and prediction of dementia. Alzheimers Dement 2006;2: 36–42.

7 Kantarci K, Jack C: Neuroimaging in Alzheimer's disease: an evidenced-based review. Neuroimaging Clin N Am 2003;13:197–209.

8 Jack CR, Knopman DS, Jagust WJ, et al: Hypothetical model of dynamic biomarkers of the Alzheimer's pathological cascade. Lancet Neurol 2010;9:119–128.

9 Jack C, Albert M, Knopman D, McKhann G, Sperling R, Carrillo M, et al: Introduction to the revised criteria for Alzheimer's disease: National Institute on Aging – Alzheimer's Association Workgroups. Alzheimers Dement 2011;7:257–262.

10 Dubois B, Feldman HH, Jacova C, DeKosky S, Barberger-Gateau P, Cummings J, et al: Research criteria for the diagnosis of Alzheimer's disease: revising the NINCDS-ADRDA criteria. Lancet Neurol 2007;6:734–746.

11 Dubois B, Feldman HH, Jacova C, Cummings J, DeKosky S, Barberger-Gateau P, et al: Revising the definition of Alzheimer's disease: a new lexicon. Lancet Neurol 2010;9:1118–1127.

Prof. Marilyn Albert, PhD
Department of Neurology, Johns Hopkins School of Medicine
1620 McElderry Street, Reed Hall East-2
Baltimore, MD 21205 (USA)
Tel. +1 410 614 3040, E-Mail malbert9@jhmi.edu

Hampel H, Carrillo MC (eds): Alzheimer's Disease – Modernizing Concept, Biological Diagnosis and Therapy.
Adv Biol Psychiatry. Basel, Karger, 2012, vol 28, pp 122–167

Pharmacological Treatment of Alzheimer's Disease

Lon S. Schneider

University of Southern California Keck School of Medicine, Los Angeles, Calif., USA

Abstract

The first section of this chapter details current, marketed pharmacological treatments for Alzheimer's disease, the extent of efficacy, adverse effects, and the controversies surrounding their use, including when to treat, for how long, expected effects, and long-term adversity. The second part overviews the rationale, development, and results of current prevention trials, all with marketed compounds, and their implications for planning future prevention trials. The third part assesses current therapy development, including drugs in phase 1, 2, and 3 clinical trials, their potential activity or inactivity, challenges of developing new treatments, including challenges of the clinical development programs, clinical trials design, and the role of proposed new diagnostic criteria in clinical trial design. Some drugs recently and currently in development are reviewed, their limitations and challenges, and what it means from a research perspective when a drug fails clinical trials. Overall, there are considerable challenges to Alzheimer's disease therapeutics as the marketed drugs, although helpful for some patients, show only small clinical effects over the short term and a degree of unappreciated toxicity over the longer term. There has not been a new drug in over a decade; drug development programs aimed at new targets have been unsuccessful thus far even though they are designed to detect small effects from the new drugs. There is nevertheless basis for considerable optimism that effective interventions soon will be found.

The first part of this chapter reviews current pharmacological treatments for Alzheimer's disease (AD), the extent of efficacy, adverse effects, and the controversy surrounding them including when to treat, for how long, expected effects, and long-term adversity. The drugs include the cholinesterase inhibitors, memantine, *Ginkgo biloba* extract, cerebrolysin, and other psychotropics. No drugs for mild cognitive impairment or to prevent AD have been approved. No drug has been approved for marketing since memantine in 2002 in Europe and 2003 in the USA.

Information on drugs is provided for general purposes only and not relied on for prescribing. Before prescribing any of the drugs discussed the physician should be knowledgeable about the full prescribing information that can be obtained from the manufacturers.

The second part discusses recent prevention trials as special cases in drug development whereby already marketed drugs and nutraceuticals are aggrandized for the purpose of preventing AD, dementia or cognitive impairment, and what can be learned from them.

The third part assesses current therapy development, including drugs in phase 1, 2, and 3 clinical trials; the challenges of developing new treatments, including challenges of clinical development and trials design, and discusses the role of proposed new diagnostic criteria in drug development [1]. This includes a summary of ongoing clinical trials, their limitations and challenges, and what it means from a research perspective when a drug fails clinical trials. As of 2008 there had been at least 172 drugs in development, and nearly all had failed by phase 2 or 3 [2]. There has not been a marketed drug for AD since 2003. It appears that events in earlier phases of drug development do not predict later ones.

Part 1: Current Treatments

Five cholinesterase inhibitors, memantine, the *G. biloba* extract, EGb 761, and cerebrolysin have some level of marketing approval for the treatment of AD at least in North America, Europe, Australia, New Zealand, or Japan. Risperidone has some level of approval in some countries in Europe for treating agitation in AD, while all antipsychotics are specifically not approved for AD in the USA.

The cholinesterase inhibitors approved for marketing in the USA are tacrine, donepezil, rivastigmine, and galantamine, approved in 1993, 1996, 2000, and 2001, respectively. At least one of the cholinesterase inhibitors, and generally at least two, are approved or on the national formularies of most countries in Europe. Memantine was approved by the European Medicines Agency (EMA) in 2002 and the Food and Drug Administration (FDA) in 2003, and is generally indicated for moderately severe to severe AD, although this varies a bit across countries. Tacrine, the first drug approved for AD in the modern era, is very rarely used if at all, and not marketed. *G. biloba* extract EGb 761 is approved or on the national formulary in Germany and France for AD, and is broadly available without prescription throughout the world.

The Cholinergic Hypothesis and Cholinesterase Inhibitors

Cholinergic treatments for AD are based on the so-called cholinergic hypothesis of dementia treatment [3]. The hypothesis implies that if cholinergic deficits are responsible for cognitive or behavioral changes in patients with dementia, then enhancing central cholinergic function will improve cognitive and behavioral function. Supporting the cholinergic hypothesis are observations of marked decline of the cholinergic basalocortical projections; marked loss of cholinergic cell bodies in the nucleus basalis; reduced activity of choline acetyltransferase that is needed for acetylcholine synthesis; correlations between choline acetyltransferase or nucleus basalis neuron loss and neuritic, β-amyloid peptide-containing plaques, and correlation of the cholinergic deficits with decline in cognitive test performance.

Despite the fact that other neuronal systems are affected in AD, including pronounced declines in the noradrenergic projections from locus ceruleus to the cortex, alterations in serotonin subtype receptors (5-HT family receptors), general changes in glutamate systems, and research focus on what are considered more fundamental pathological processes affecting neurodegeneration, including the β-amyloid cascade and tau protein metabolism, and inflammatory responses, intervention with cholinesterase inhibitors remains the only consistently demonstrated treatment for mild to moderate AD although their efficacy leaves much to be desired (see below).

Three main cholinergic approaches to AD treatment have included the use of precursor loading, direct cholinergic agonists, and cholinesterase inhibition. The first two have not been proven clinically effective and are discussed elsewhere [4]. Barriers to the successful development of muscarinic agonists involve the difficulty in finding a therapeutic with clear M_1 subtype agonism and few adverse effects [5]. The potential for nicotinic neuronal receptor modulators as AD therapeutics is discussed below.

Donepezil

Donepezil is a long-acting, piperidine-based, selective and reversible acetylcholinesterase inhibitor. Two phase 3 clinical trials provided pivotal evidence for the drug's FDA approval in late 1996. Additional randomized clinical trials include a trial of 6 months' duration, a Nordic study of 12 months, a study of nursing home patients, a 6-month trial in which a large proportion of patients were more severely impaired than in previous trials, and an assessment of donepezil's activities of daily living, as reviewed in a Cochrane Review [6].

One non-industry-sponsored, randomized, placebo-controlled trial followed patients over several years, reporting very modest cognitive effects over 2 years but no significant effects on daily function, nursing home placement, or health economic measures [7].

Pharmacokinetics and Drug Interactions

Oral bioavailability approaches 100%, with a time to peak concentration, T_{max}, in 4 h and linear pharmacokinetics. It is extensively bound to plasma proteins, steady state is reached in approximately 2 weeks. Clearance is slow, with a long elimination half-life of 70 h. It is both excreted unchanged in the urine and extensively metabolized in the liver by CYP2D6 and CYP3A4 to active and inactive metabolites. A 23-mg extended-release formulation has been marketed recently, intended for more severe AD patients already receiving 10 mg/day. It has a slower absorption and prolonged T_{max}.

Inhibitors of CYP3A4 and CYP2D6 (such as ketoconazole and quinidine, respectively) and inducers of CYP2D6 and CYP3A4 (such as phenytoin, carbamazepine, dexamethasone, and phenobarbital) could inhibit metabolism or increase elimination of donepezil. However, this may not be clinically significant in view of the drug's high protein binding and very low therapeutic concentrations from 2–12 ng/ml.

Rivastigmine

Rivastigmine is a pseudoirreversible cholinesterase inhibitor that is selective for acetylcholinesterase and butyrylcholinesterase. Four phase 3 randomized, placebo-controlled, 26-week-long clinical trials were completed of similar design but differing mainly in dosing methods. Two have been published, and the results of the other two have been included in secondary reports [8, 9].

In the published trials, doses were titrated weekly over 7 weeks to one of two dosage ranges, 1–4 or 6–12 mg/day, and dose decreases were not permitted, possibly contributing to lesser tolerability and seemingly more side effects during these stages of treatment [10, 11].

A transdermal patch formulation has been marketed based on a randomized placebo-controlled trial comparing a 17.4-mg patch, 9.5-mg patch, and 6 mg of orally administered rivastigmine twice per day in 1,195 moderately severe AD patients – i.e. Mini-Mental State Examination (MMSE) 10–20 – over 6 months showing efficacy on the Alzheimer's Disease Assessment Scale – Cognitive subscale (ADAS-cog) and global ratings, and fewer adverse events than with rivastigmine capsules [12].

Pharmacokinetics and Drug Interactions

Rivastigmine is well absorbed and has very little protein binding. Although the elimination half-life is <2 h, enzyme inhibition lasts about 9 h. The drug is not metabolized by the hepatic microsomal system. Rather, after binding to acetylcholinesterase, the carbamate portion of rivastigmine is slowly hydrolyzed (hence the 'pseudoirreversibility'), conjugated, and excreted by the kidneys. The pharmacokinetics of the transdermal patch shows a time to maximum concentration of 8–16 h and a 3-hour elimination half-life after the patch is removed. Because its metabolism is extrahepatic and does not depend on liver cytochrome enzymes, it is unlikely to have significant pharmacokinetic interactions.

Galantamine

Galantamine, an alkaloid originally found in Amaryllidaceae (*Galanthus woronowi*, the Caucasian snowdrop), is a reversible, competitive acetylcholinesterase inhibitor with relatively less butyrylcholinesterase activity. A competitive inhibitor would compete with acetylcholine for acetylcholinesterase binding, and thus inhibition would be dependent on intrasynaptic acetylcholine concentration. Competitive inhibitors should be less likely to be active in brain areas that have remaining high acetylcholine levels and more active in other areas. Galantamine's other mechanistic characteristic is as an allosteric modulator of nicotinic receptor sites, possibly enhancing cholinergic transmission by presynaptic nicotinic stimulation.

At least four 3- and 6-month long, randomized, placebo-controlled, clinical trials involving more than 2,400 subjects have been published, generally showing efficacious doses between 8 and 16 mg b.i.d., with fewer adverse effects at the low dose. A Cochrane Review concluded that galantamine shows consistent positive effects of 3–6 months' duration without a response improvement with doses >16 mg/day and that its safety profile is similar to that of other cholinesterase inhibitors with respect to gastrointestinal symptoms [13].

Pharmacokinetics and Drug Interactions

Galantamine is well absorbed with approximately 90% bioavailability, little protein binding, with peak concentrations in approximately 1 h, an elimination half-life of approximately 7 h, and linear pharmacokinetics. It is metabolized by CYP2D6 and CYP3A4, glucuronidated, and excreted in the urine. A sustained-release form is marketed as well.

Drug Interactions

Drugs that are potent inhibitors for CYP2D6 or CYP3A4 may increase the availability and plasma levels of galantamine, for example, paroxetine and ketoconazole, respectively. Clearance of galantamine is decreased approximately 30% with administration of CYP2D6 inhibitors, such as amitriptyline, fluoxetine, and quinidine, among other drugs. Galantamine itself does not inhibit CYP3A4 or CYP2D6 and has little effect on the metabolism of other drugs.

Adverse Effects of Cholinesterase Inhibitors

Adverse effects are similar across the three cholinesterase inhibitors. Nausea, diarrhea, vomiting, and weight loss are most common (table 1).

Muscle cramps are common with donepezil. Significant cholinergic effects may occur in approximately one quarter of patients and are often related to initiation and titration of medication. Reducing the dose temporarily and then reinstituting using a slower titration may be helpful. Many patients tend to become tolerant to the adverse events. Weight loss, fatigue, and anorexia may be insidious, lead to profound problems, and not be recognized as adverse effects of these drugs. Few trials have directly compared cholinesterase inhibitors between themselves with respect to adverse events [14].

An increase in anorexia is a consistent finding across trials and may be dose-related. The incidence varies from approximately 8 to 25% at higher doses of cholinesterase inhibitors compared with 3–10% in placebo patients. With respect to weight loss, the proportion of patients losing greater than 7% of their weight ranges from 10 to 24% in patients taking higher doses compared to 2–10% of placebo-treated patients in trials reporting the statistic.

Table 1. Summary of adverse event data from placebo-controlled, randomized clinical trials [modified with permission from 14]

Drug	Adverse events (% vs. placebo)
Donepezil	Nausea, diarrhea, insomnia, vomiting, muscle cramps, fatigue, anorexia, dizziness, abdominal pain, myasthenia, rhinitis, weight loss, anxiety, syncope (2 vs. 1%)
Rivastigmine	Nausea, vomiting, anorexia, dizziness, abdominal pain, diarrhea, malaise, fatigue, asthenia, headache, sweating, weight loss, somnolence, syncope (3 vs. 2%). Rarely, severe vomiting with esophageal rupture
Galantamine	Nausea, vomiting, diarrhea, anorexia, weight loss, abdominal pain, dizziness, tremor, syncope (2 vs. 1%)

Methods for obtaining and reporting adverse events vary among trials, making it difficult to determine relative rates of adverse events when comparing drugs.
Cholinergic adverse events generally occur early in the course of treatment related to initiating or increasing medication. Adverse events tend to be mild and self-limited. Medications should be temporarily stopped and restarted at lowest doses.

Some general precautions with cholinesterase inhibitors listed in the prescribing information for the drugs are that by increasing central and peripheral cholinergic stimulation, they may increase gastric acid secretion, increasing the risk for gastrointestinal bleeding especially in patients with ulcer disease or those taking antiinflammatories; produce bradycardia, especially in patients with sick sinus or other supraventricular conduction delay, leading to syncope, falls, and possible injury; exacerbate obstructive pulmonary disease including asthma; cause urinary outflow obstruction; increase risk of seizures, and prolong the effects of succinylcholine-type muscle relaxants.

Long-term safety has not been well studied. Using medical and prescription records, patients on cholinesterase inhibitors (mainly donepezil) were hospitalized for syncope nearly twice as often as people with dementia who did not receive these drugs. Experiencing a slowed heart rate was 69% more common among cholinesterase inhibitor users. In addition, people taking the dementia drugs had a 49% increased chance of having permanent pacemakers implanted and an 18% increase risk for hip fractures [15]. The low absolute incidence of these events is about 2% of treated patients per year and may not be fully appreciated in a clinical context. Yet, it implies that for every 50–100 patients treated for 1 year, 1 will be hospitalized because of syncope due to the drug.

Effectiveness of Cholinesterase Inhibitors

The three cholinesterase inhibitors generally show efficacy for mild to moderate AD, and donepezil for severe AD as well (see below). Efficacy, however, was demonstrated mainly in 3- and 6-month long trials. The relevant clinical question however is not

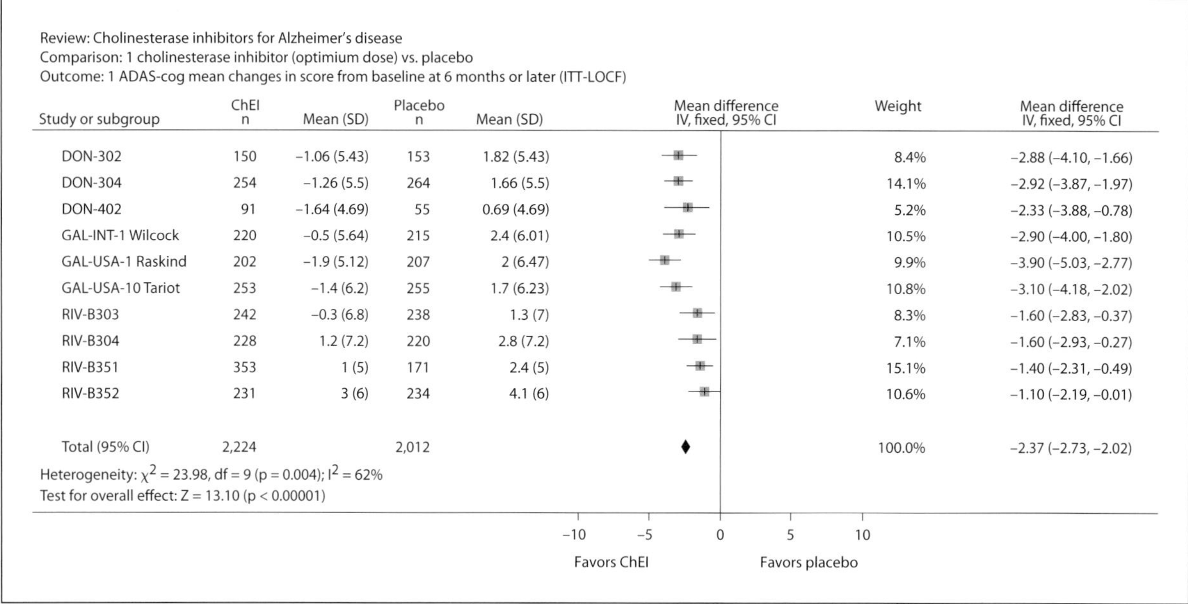

Review: Cholinesterase inhibitors for Alzheimer's disease
Comparison: 1 cholinesterase inhibitor (optimium dose) vs. placebo
Outcome: 1 ADAS-cog mean changes in score from baseline at 6 months or later (ITT-LOCF)

Study or subgroup	ChEI n	Mean (SD)	Placebo n	Mean (SD)	Weight	Mean difference IV, fixed, 95% CI
DON-302	150	−1.06 (5.43)	153	1.82 (5.43)	8.4%	−2.88 (−4.10, −1.66)
DON-304	254	−1.26 (5.5)	264	1.66 (5.5)	14.1%	−2.92 (−3.87, −1.97)
DON-402	91	−1.64 (4.69)	55	0.69 (4.69)	5.2%	−2.33 (−3.88, −0.78)
GAL-INT-1 Wilcock	220	−0.5 (5.64)	215	2.4 (6.01)	10.5%	−2.90 (−4.00, −1.80)
GAL-USA-1 Raskind	202	−1.9 (5.12)	207	2 (6.47)	9.9%	−3.90 (−5.03, −2.77)
GAL-USA-10 Tariot	253	−1.4 (6.2)	255	1.7 (6.23)	10.8%	−3.10 (−4.18, −2.02)
RIV-B303	242	−0.3 (6.8)	238	1.3 (7)	8.3%	−1.60 (−2.83, −0.37)
RIV-B304	228	1.2 (7.2)	220	2.8 (7.2)	7.1%	−1.60 (−2.93, −0.27)
RIV-B351	353	1 (5)	171	2.4 (5)	15.1%	−1.40 (−2.31, −0.49)
RIV-B352	231	3 (6)	234	4.1 (6)	10.6%	−1.10 (−2.19, −0.01)
Total (95% CI)	2,224		2,012		100.0%	−2.37 (−2.73, −2.02)

Heterogeneity: $\chi^2 = 23.98$, df = 9 ($p = 0.004$); $I^2 = 62\%$
Test for overall effect: Z = 13.10 ($p < 0.00001$)

Fig. 1. Cholinesterase inhibitors, optimum dose versus placebo. The table and figure show the mean drug-placebo difference on the ADAS-cog from several clinical trials along with 95% CI widths displayed as horizontal lines. The overall mean effect is 2.37 points with 95% CI −2.73 to −2.02 [reproduced from 16, with permission].

efficacy per se but effectiveness, a question of whether the efficacy seen in clinical trials has an overall meaningful clinical effect.

Cognitive function was nearly always assessed using the ADAS-cog, with a few trials using the MMSE, as the primary outcome. As co-primary or secondary outcomes the trials used a clinician's global assessment of change or a daily activities inventory. Quality of life and behavioral symptoms as rated by caregivers were also assessed in many studies.

Despite the differences in mechanism of action, as described above, there is no evidence for any efficacy differences between the three cholinesterase inhibitors. Efficacy can be fairly described as, first, rather consistent from trial to trial and, second, rather modest, especially in terms of both the mean magnitudes of change compared with baseline function and compared to change with placebo treatment. As presented in a Cochrane Review [16], it can be seen that the drugs are associated with an overall mean 2.4 effect, and range of means from 1.1 to 3.9 points, over placebo on the ADAS-cog outcomes with no differences between them. Moreover, absolute mean improvement with treatment over 6 months ranges from about no change or worsening to about 1.9 points improvement on the ADAS-cog points (fig. 1).

The 2.4-point difference represents about 10% change on the ADAS-cog scale, within the bounds of the test-retest reliability of the outcomes, and does not necessarily represent a change in the patient's clinical state. Although clearly there are some patients who improved substantially with cholinesterase inhibitors, there are also some who worsened to a substantially greater extent than placebo-treated patients as well. It is frustrating that precisely as a consequence of the design of the clinical trials and the modest therapeutic effect it is not possible to identify clear treatment responders or those who will respond to treatment prior to treatment [16].

Although there may be fewer acute adverse effects associated with donepezil compared to the other drugs, the differences are less so with the use of sustained-release galantamine or transdermal rivastigmine. Intolerability may have more to do with initial titration than with longer term treatment, at least with respect to nausea and vomiting. Markedly affecting overall effectiveness is that withdrawals due to adverse events are significantly higher for treatment with all three cholinesterase inhibitors than placebo. Thus the use of cholinesterase inhibitors involves balancing the modest expectations for help against the expectations for intolerance and adverse effects due to the drugs.

Cholinesterase Inhibitors for Severe Alzheimer's Disease

Donepezil is the only cholinesterase inhibitor labeled for severe AD in addition to mild to moderate, the indications for rivastigmine and galantamine. The efficacy evidence for donepezil in severe AD is based on three 6-month, randomized, placebo-controlled clinical trials in patients with MMSE scores of ≤10 and in which a different cognitive outcome, the Severe Impairment Battery (SIB) is used in place of the ADAS-cog because patients this severe cannot reliably be assessed on the latter's subtests [17]. Effects here, too, were of the order of several points on a scale ranging from 100 to 0. In these trials intended for regulatory approval, patients had not been previously exposed to cholinesterase inhibitors. Thus the efficacy test is not particularly generalizable to clinical practice, at least in the USA, as it would be rare to find severe AD patients who had not been exposed to cholinesterase inhibitors previously or are not currently taking them.

A 23-mg extended-release formulation of donepezil has been marketed, intended to be used after a patient has been treated with 10 mg/day for at least 3 months, has reached a severe level of cognitive impairment, and when there is uncertainty on whether the patient is benefiting from the 10-mg dose. In the only clinical trial to document its efficacy, 1,467 AD patients with MMSE scores from 20 to 0 and who had been treated with donepezil 10 mg/day were randomized to 23 mg or to continue their 10-mg dose for 6 months [18]. A significant 2.2 drug-placebo difference on the SIB was not supported by efficacy on the clinician's global assessments, and dropouts were greater with the 23-mg dose than the continuing 10-mg dose, 30 vs. 18%. A post-hoc analysis in the more severe patients suggested a somewhat greater effect on the SIB, about 3.1 points, and significance on the global assessment of 0.09 points. The formulation and marketing approval are controversial, however, because the effects are marginal, and it was approved by FDA despite recommendations for non-approval by two FDA officers. Moreover, a citizen's watchdog group expressed concern about the added risks for the higher dose in relationship to any added benefits. As donepezil goes off patent in 2012, this 23-mg extended-release formulation will be the only 'Aricept' branded drug, selling at a premium over the generics and heavily promoted.

Cholinesterase Inhibitors for Mild Cognitive Impairment
Several trials assessed efficacy and safety for cholinesterase inhibitors for mild cognitive impairment. These randomized, placebo-controlled trials used the typical doses for AD, tended to be long-term, from 2 to 4 years, and used the onset of AD as main endpoints. Two galantamine trials of 2 years' duration [19] – one rivastigmine trial of 4 years [20] and one donepezil vs. vitamin E trial over 3 years [21] – constitute the corpus of the trials, in addition to a 6-month donepezil trial [22]. All showed negative results on their primary outcomes and an excess in adverse events, and have been summarized in at least three systematic reviews [13, 23, 24]. One pointed out that uncertainty regarding the definitions of MCI casted doubts on the validity of such trials [23]. The galantamine trials were particularly concerning for an excess in deaths. Also, of note, the two donepezil trials used MCI definitions that are nearly identical with the currently proposed MCI due to AD diagnosis and very close to the definition of AD itself. In these trials nominal drug-placebo differences on the ADAS-cog were observed at some points. The effects, however, were not sustained, and seemed more evident in the overlapping post-hoc subgroups of the relatively more severely impaired mild cognitive impairment patients at baseline and in ApoE ε4 carriers, over the first year or so.

Dosage and Use
The preparations and dosages of cholinesterase inhibitors are listed in table 2. With all cholinesterase inhibitors, raising the dose too soon increases risks for cholinergic adverse events.

Donepezil is started at 5 mg/day and can be increased to 10 mg/day after 4 weeks. Effective doses are 5 or 10 mg/day; 10 mg tends to be somewhat more effective than 5 mg when group data from various clinical trials are compared. The efficacy or appropriateness of higher doses has not been established. The sustained-release 23-mg donepezil formulation is restricted for use in severe AD after patients have been treated with 10 mg/day for a period of time.

The starting dose of rivastigmine is 1.5 mg b.i.d., taken with meals. If this dose is well tolerated after a minimum of 2 weeks of treatment, it may be increased to 3 mg b.i.d. Subsequent increases to 4.5 mg b.i.d. and then 6 mg b.i.d. could be based on tolerability and considered after a minimum of 2 weeks of treatment. The initial transdermal patch is applied at 4.6 mg (5 cm^2) per day and then increased to one 9.5-mg (10 cm^2) patch per day after 4 weeks with maintenance dose of either 4.6 or 9.5 mg/day.

Initial dosing for galantamine is 4 mg b.i.d. and should be raised to 8 mg b.i.d. after 4 weeks. For patients who are tolerating medication but not responding, the dose can be raised to 12 mg b.i.d. after another 4 weeks. The initial dose of the extended-

Table 2. Dosages of marketed cholinesterase inhibitors [adapted with permission from 4]

Drug	How supplied	Initial dosage	Maintenance dosage	Comments
Donepezil	Tablets: 5 and 10 mg Orally disintegrating tablets: 5 and 10 mg	5 mg q.d.	5–10 mg q.d.	10 mg may be somewhat more efficacious than 5 mg q.d. in some trials
Donepezil-sustained release	23 mg sustained release	23 mg q.d. (see comment)	23 mg q.d.	For use only in patients with severe AD who have been maintained on 10 mg q.d.
Rivastigmine	Capsules: 1.5, 3, 4.5, and 6 mg	1.5 mg b.i.d.	3, 4.5, or 6 mg b.i.d.	Effective dosage range 3–6 mg b.i.d. Doses of 4.5 mg b.i.d. may be most optimal. May be taken with food
Rivastigmine transdermal	Transdermal patch: 4.6 and 9.5 mg/24 h	4.6 mg/24 h	9.5 mg/24 h	Avoids first-pass hepatic metabolic effect, better tolerated than oral
Galantamine	Tablets: 4, 8, and 12 mg Oral solution: 4 mg/ml	4 mg b.i.d. 8 mg q.d.	8 or 12 mg b.i.d.	Effective dosage range is 16–24 mg/day; 16 mg/day is the modal optimal dose
Galantamine-extended release	Extended-release capsules: 8, 16, and 24 mg	8 mg q.d.	16 or 24 mg q.d.	Effective dosage range is 16–24 mg/day

Initial and subsequent dosages should be maintained for at least 2, preferably 4–6, weeks before increasing. Adverse events may occur with upward dosage titration. For prescribing information see: http://www.aricept.com/productinfo.htm; http://www.pharma.us.novartis.com/product/pi/pdf/exelon.pdf; http://razadyneer.com/razadyneer/pages/pdf/razadyne_er.pdf.
Tacrine is very rarely used currently, if at all, and is not included here.

release formulation is 8 mg/day and can be raised to 16 mg/day after 4 weeks with the option to increase to 24 mg/day.

Memantine

Memantine is a moderate-affinity, uncompetitive N-methyl-D-aspartate (NMDA) receptor antagonist, that had been available in Germany for the treatment of spasticity, Parkinson's disease, and other neurological disorders since 1982 but not used as such. A rationale for its use is that overstimulation of NMDA receptors may occur in patients with AD and the drug protects against the effects of glutamate and calcium-mediated neurotoxicity.

Memantine was approved for moderate to severe AD – meaning patients with MMSE scores <15 – by the FDA in late 2003 on the basis of the following trials: a placebo-controlled, 6-month trial published in 2000 [25] in which cholinesterase inhibitors were not allowed, and another 6-month trial published in 2004 in which all patients had been taking donepezil for at least 6 months (actually, over 2 years on average) prior to being randomized to memantine or placebo [26]. A third, moderate to severe AD trial did not favor memantine, however [27].

The clinical trials were similar in design to those used with cholinesterase inhibitors but included patients with moderate to severe AD – i.e. MMSE <15 – as compared to mild to moderate AD which would be operationalized as the MMSE range from 10 or 12 to 24 or 26.

The 6-month long trials used the SIB (Severe Impairment Battery), CIBIC+ (Clinician's Interview-Based Impression of Change with Caregiver Input) or ADCS-ADL (Alzheimer's Disease Cooperative Study Activities of Daily Living) as primary outcomes, and ADL or behavior as secondary outcomes. The SIB is used as the cognitive test because the patients are too severe to perform on the ADAS-cog. The ADL scale is also modified to exclude daily tasks that severe dementia patients cannot do. Nevertheless, the *statistical* magnitudes of benefit from treatment with memantine in moderate to severe AD licensing trials is similar to modest effect sizes that are observed for cholinesterase inhibitors in mild to moderate AD trials.

A 12-week clinical trial of memantine for patients with dementia, i.e. both AD and vascular dementia, residing in nursing homes published in 1999 [28] showed functional improvement with memantine; and along with the 6-month monotherapy trial above supported its earlier approval for moderate to severe AD by the EMA in 2002.

In mild to moderate AD clinical trials, very similar in design to the cholinesterase inhibitor trials, one memantine trial demonstrated statistically significant improvement on the ADAS-cog, CIBIC+, and behavior, although the results depended on the statistical analysis method used [29]. Two other, similarly designed trials, also intended to gain marketing approval, did not show significant drug-placebo differences [30, 31]. In pooled analyses of these trials, memantine showed no efficacy for mild AD – i.e. MMSE >19 – and less efficacy in the moderate AD range – i.e. MMSE 15–19 – than in lower ranges [32]. Thus, it is not approved by the FDA for patients with mild AD, meaning MMSE scores >14 [33]. European authorities, however, extended the range of approval for use of memantine to patients with MMSE scores up to 19.

Recently a 28-mg sustained-release oral preparation has been reported to be effective in a 6-month, placebo-controlled, severe AD trial, and involving 677 patients with MMSE scores from 3–12. There were statistically significant differences in outcomes favoring memantine on the SIB (2.7 points difference from placebo), CGIC (Clinical Global Impression of Change, about 0.3 points), and on a behavior rating scale, but no differences on activities of daily living. Adverse events were reported as similar to placebo. Trial results, however, have not been published. The sponsors report that the formulation has been approved by the FDA (http://www.medscape.com/viewarticle/581440, http://www.frx.com/news/PressRelease.

aspx? ID=1181404). Although it has not been publicly stated why the formulation has not yet been marketed, the delay may involve commercial considerations in switching to this new, branded formulation prior to the current memantine formulation becoming generic.

The corpus of memantine trials are summarized in a Cochrane Review that concluded that memantine had a small beneficial effect in moderate to severe AD and was well tolerated [34].

Mechanism of Action

Although its actual therapeutic mechanism of action is unknown, memantine is hypothesized to work as an open-channel NMDA receptor antagonist without appreciable pharmacological activity until increased glutamate levels trigger the receptor and cause the ion channel to open. The drug is thought to then enter the channel, preventing calcium influx, depolarization, and hyperactivation of the neuron. Its action, therefore, might depend on intrasynaptic glutamate concentration in that as glutamate increases and causes channels to open, memantine may block more, and as glutamate decreases, it may block less, allowing greater neuronal activity. Thus another perspective is that memantine modulates glutamatergic activity.

The above hypothesized mechanism represents a putative neuroprotective effect, and it is speculative how it explains memantine's effect in 6-month trials as this clinical effect is usually described as 'symptomatic.' The drug may have other effects, however, such as on long-term potentiation that may correlate with a short-term effect on memory [35].

Some associations between NMDA and glutamate receptors and AD include that glutamate transporter is downregulated in AD, leading to increased synaptic glutamate; glutamate receptor activity and β-amyloid inhibits the uptake of glutamate and further enhances glutamate release; the excitotoxicity may increase levels of amyloid precursor protein (APP), and increased NMDA receptor activity may increase tau protein phosphorylation. Other preclinical studies suggest that memantine may reduce Aβ plaque deposition and tau hyperphosphorylation [36]. Notably, the potential to moderate these effects with memantine has not been tested clinically, perhaps because this would require long-term and large trials, and the known clinical effects of memantine remain modest and mainly in those with severe AD.

Pharmacokinetics

Absorption of memantine is not affected by food; bioavailability approaches 100%; plasma protein binding is about 45%, plasma concentration maximum, C_{max}, occurs 3–7 h after oral intake; it is widely distributed and readily diffuses across the blood-brain barrier; there is minimum hepatic metabolism, and it is mostly excreted unchanged in the urine; elimination half-life is 60–80 h.

Pharmacokinetics of extended-release memantine is reported as T_{max} of 12 h, with a lower C_{max} and an elimination half-life of 41–74 h.

Adverse Effects of Memantine

Adverse events reported with memantine therapy include headache, dizziness, confusion, somnolence, and infrequent hallucinations. Interestingly, agitation occurred less often in memantine treatment groups than in placebo groups. Also gastrointestinal symptoms are decreased in frequency, with diarrhea or fecal incontinence occurring half as frequently in the memantine than in a placebo group, suggesting that the actions of memantine may mitigate some of the cholinergic effects of donepezil (although this has not been clearly studied). No adverse drug interactions have been reported with patients receiving cholinesterase inhibitors in addition to memantine, and the drugs appear safe when used together.

Dosage and Use

Memantine is titrated at 5 mg/day for 1 week, then 5 mg b.i.d. the next week; 10 mg in the morning and 5 mg later in the day for the third week, and 10 mg b.i.d. from then on. The reason for this titration regimen – especially in light of memantine's long half-life and high degree of tolerability – is because it was used in the trials that supported its licensing, and other regimens have not been tested. Notwithstanding the long half-life, the pharmaceutical manufacturers developed a 28-mg sustained-release version, to be administered once per day (and still with a 3-week titration period, but with 7-, 14-, and 28-mg forms), that will be marketed when the current memantine formulation becomes available as a generic in the near future.

Memantine is given either alone or in addition to a cholinesterase inhibitor, often after the latter has been used for a long period. Some clinicians initiate memantine along with or soon after a cholinesterase inhibitor in patients with mild AD. As with cholinesterase inhibitors, its duration of effectiveness is not known beyond the length of the clinical trials, and it tends to be prescribed for indefinite periods. Some open-label observations from clinic cohorts suggest that combination therapy with a cholinesterase inhibitor and memantine may ameliorate the course of AD [37, 38], while other observations from clinical trial extensions and ADNI suggest that the effect of memantine may wane [39, 40].

Observations in the course of doubled-blind placebo-controlled trials suggested that patients on therapy decline at equal rates when compared to patients not receiving therapy [41]. The practical issue of how long to treat is particularly challenging in patients with severe dementia and poor quality of life. Severely impaired patients are often treated until death.

Other Marketed Compounds

Some other compounds are licensed for AD or listed on national formularies for reimbursement. They include *G. biloba* standardized extract and cerebrolysin, and so are discussed briefly. *Cerebrolysin* is a preparation of peptides and amino acids derived from porcine brain proteins that preclinically has neurotrophic actions. It is widely available in Europe and Asia, and is administered intravenously 5 days per week for 4-week periods with effects considered to last up to 3 months. Several 6-month, placebo-controlled trials have been conducted in mild to moderate AD patients that showed equivocal outcomes and will not be discussed further [for review, see 42].

Ginkgo biloba

G. biloba leaves and extracts are widely used over-the-counter preparations marketed in the USA as food supplements for which health claims are not permitted and insurance does not reimburse [43]. A specific standardized extract, EGb 761, is approved for the national formularies in some countries, Germany and France, in particular. This is standardized to contain two major constituents: 22–27% flavonoids and 5–7% terpene lactones (ginkgolides and bilobalide).

The flavonoids are active as antioxidants and appear neuroprotective. Ginkgolide B is an antagonist of the platelet-activating factor receptor. Ginkgolides A and J variously inhibit hippocampal neuron dysfunction and neuronal cell death caused by $A\beta_{42}$. Ginkgolides A and J decrease $A\beta_{42}$-induced pathological behaviors, enhance neurogenesis in animal models of AD, and inhibit Aβ aggregation, providing considerable rationale for *G. biloba* extracts as potential treatments for AD. None of this has been demonstrated in humans, however.

Trials in older and younger adults who do not have cognitive impairment show mixed results at best [43]. One meta-analysis of eight trials did not find evidence for cognitive benefits with *G. biloba* in non-cognitively impaired participants younger than 60 years treated for up to 13 weeks. Two placebo-controlled trials reported contradictory effects in non-cognitively impaired older adults, and the magnitude of the cognitive effects were small in the positive trial.

A Cochrane systematic review that included 35 clinical trials reported inconsistent evidence that *G. biloba* had clinically significant benefits for dementia or cognitive impairment [44]. One 6-month trial in mild to moderate AD conducted with the hope of gaining FDA approval failed to demonstrate efficacy [45], as did another 6-month trial performed at British primary care sites with 120-mg/day doses of EGb 761 [46].

Conclusions of the German Institute for Quality and Efficiency in Health Care (IQWiG). This group found that EGb 761 at 240 mg/day might improve activities of daily living, and there were 'indications of a benefit' for cognitive function, general psychopathological symptoms, and improvement in caregiver emotional stress. The benefit, however, was only present in dementia patients with psychopathological symptoms. Their conclusions were 'based on very heterogeneous results', and 'were strongly affected by two studies conducted in an Eastern European healthcare setting' (https://www.iqwig.de/index.948.en.html).

G. biloba Prevention Trials. Of particular interest for drug development, two large prevention trials involving approximately 2,700–3,100 patients followed over 5–7 years [47, 48; Vellas et al., unpubl. data], and a small trial in patients over the age of 85 [49] did not yield significant results (see discussion below). Thus, there is very little evidence for the efficacy of EGb 761 either for improving symptoms or preventing AD.

Controversies Involving Current Treatments

The limited efficacy and differential adverse effects of the available drugs, combined with the dire straits and unmet needs of many AD patients and their families, has, not surprisingly, led to substantial controversy about the use and usefulness of current marketed treatments. The overarching controversy transcends the polar views that (1) the drugs are effective and the standard of care for AD, or (2) they are ineffective and not worthwhile in terms of cost and adversity.

A main controversy is not whether the drugs have measured efficacy, because they do consistently show significant drug-placebo differences, but whether or not they are effective enough, and whether their use yields clinically meaningful or therapeutically useful outcomes. Some argue that cholinesterase inhibitors have failed to provide substantial improvement for AD patients. This discussion is played out in the responses to a rather strident and nihilistic systematic review that concluded, 'Because of flawed methods and small clinical benefits, the scientific basis for recommendations of cholinesterase inhibitors for the treatment of Alzheimer's disease is questionable [50]' (http://www.bmj.com/content/331/7512/321.abstract/reply#bmj_el_120412).

The US Agency for Health Care Quality (AHCQ) was more tempered in its review concluding, 'Treatment of dementia with cholinesterase inhibitors and memantine can result in statistically significant but clinically marginal improvement in measures of cognition and global assessment of dementia' [51]. The UK's NICE and a health assessments group initially concluded in 2006 that the cholinesterase inhibitors can delay cognitive impairment over 6 months [52], and that they are not cost-effective [53]. They added further concerns that the real impact of the drugs is difficult to assess, that the studies discussed above were of moderate quality, the samples are not generalizable, do not account for patients with medical co-morbidity, and the large variability in outcomes means that the mean scores are difficult to interpret as well [52]. NICE is less restrictive in their revised opinion.

The issue of essential effectiveness has been further obscured by pharmaceutical marketers' advertising campaigns directed at both physicians and caregivers that, arguably, through images and music, present the drugs as more effective than they may be. Some advertisements have led to FDA warnings. For example, Eisai received a warning in 2009 for misleading television advertisements for donepezil, suggesting that the drug was more effective than it is (http://www.fda.gov/downloads/Drugs/GuidanceComplianceRegulatoryInformation/EnforcementActivitiesbyFDA/WarningLettersandNoticeofViolationLetterstoPharmaceuticalCompanies/UCM201238.pdf).

The blockbuster status of the drugs also fuel demand. Annual sales of donepezil in the USA are approximately USD 2 billion and of memantine USD 1 billion. By

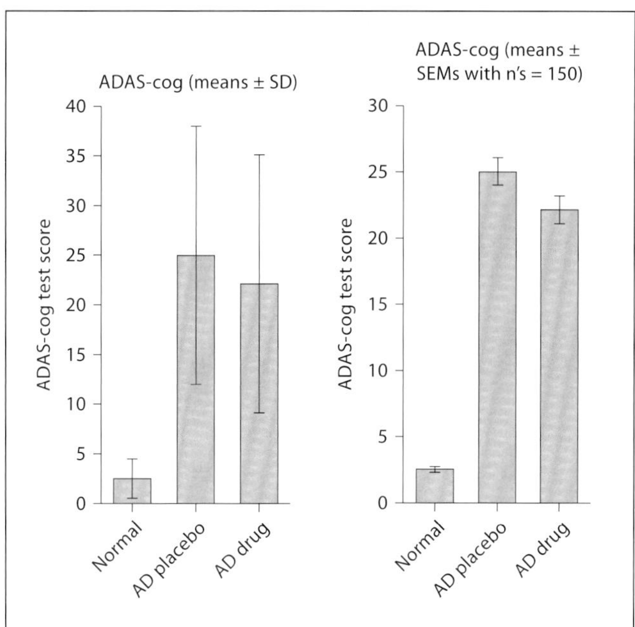

Fig. 2. Comparison of the identical effect of donepezil from a clinical trial compared to placebo with standard deviation bars (left), reflecting the distribution of outcomes, and standard error of measurement limits (right), reflecting the precision of the outcomes [reprinted from 2, with permission].

comparison, the National Institutes of Health research budget for AD in 2012 is estimated at USD 458 million (http://report.nih.gov/rcdc/categories/Default.aspx). As the costs of these drugs are substantially covered by Medicare, i.e. paid for by the US government, their cost-effectiveness is not an issue in the USA.

Efficacy vs. Effectiveness

The effectiveness controversy is illustrated in figure 2 that shows the typical 2.5- to 3-point effect on the ADAS-cog with donepezil compared to placebo. One side shows the precision of the effect as bracketed with standard error bars; the other side the distributions of the ADAS-cog outcomes as described with standard deviation bars. The narrow standard error bar indicates the rather strong statistically significant drug versus placebo effect of donepezil, and in conjunction with table 3, illustrates the robust statistically significant cognitive effect of cholinesterase inhibitors across drugs and trials. The standard deviation bars on the same study, however, first show the broad and substantial overlap in outcomes between drug- and placebo-treated patients such that very few individual can be seen to benefit, and, second show that even among those patients who improve to the greatest extent, their cognitive scores are still in the moderately severe dementia range. Together, the figure illustrates what is meant by the cognitive effects of cholinesterase inhibitors being small, and why interpreting this effect into improved health outcomes is difficult.

It is difficult to identify the individual patient who benefits from cholinesterase inhibitors or memantine as the outcomes measures used in the trials and clinically are not suitable for identifying responders. Moreover, most clinical trials have not taken patient/caregiver views into account, although one that did shows that important aspects of the treatment response are missed by current measures [54]. More could be done to parse out the effects to address why some patients benefit and others do not.

How Long to Treat?

Placebo-controlled clinical trials with marketed cholinesterase inhibitors have lasted only 6 months with only a few exceptions, and long-term effects beyond 1 year have not been studied in appropriately controlled databases. Two placebo-controlled trials suggest that in some cases efficacy could be longer than 1 year on cognitive and functional measures [7, 55]. Over the long term, as patients inevitably worsen, it becomes more difficult to determine whether individuals are benefiting. Assumptions are made that if the drugs are effective for 6 months then they will continue to be beneficial far longer, perhaps indefinitely.

In some 3- and 6-month long trials, after medication has been discontinued, patients on average return within 6 weeks to the cognitive level of the contemporaneously placebo-treated patients. This is taken as indicating an overall symptomatic effect from the medication, in that the medications create no structural effect on the course of illness, and continuous use is required to maintain benefits.

A few analyses of clinic databases or open-label extensions of clinical trials suggest that patients who tolerate and continue to take cholinesterase inhibitors over at least 1 year might experience a delay in nursing home placement compared to those who cannot tolerate or do not take them and that the further addition of memantine enhances that effect [56–59]. These observations, however, are inherently biased, subject to a 'healthy patient effect' or 'survivor effect' whereby patients who are destined to have a milder course of illness or who have active caregivers continue take these medications while others do not so that the apparent therapeutic effect is illusory, and by comparisons across clinic cohorts from different time periods before and after the advent of the drugs [60]. These observational studies, however, are inconsistent with the long-term, placebo-controlled trials of cholinesterase inhibitors for mild cognitive impairment, and with observational analyses with ADNI [61] and AIBL datasets [62], where in the latter it appears patients maintained on cholinesterase inhibitors decline faster. Better, effectiveness research on long-term treatment is needed to resolve the uncertainties.

Are There Effects from Withdrawal of Cholinesterase Inhibitors?

Discontinuation of cholinesterase inhibitors occasionally has been associated with worsening of cognition and episodes of confusion in some patients especially after abrupt withdrawal. This is evident in clinical trials when the drugs are stopped after a fixed, period of 12 weeks open-label treatment and patients are then randomized to continuing drug or to placebo [63], and when they are stopped at the end of some 6-month trials. Behavioral effects and confusion do not appear common in clinical practice when the drugs are stopped, as they frequently are, because they appear not to be working or there is uncertainty about their efficacy. In clinical practice, about 31% discontinue rivastigmine or donepezil within 60 days, the rest continued for just short of 11 months, and about 19–23% were continuing to take the drugs for over 1 year [64]. It is generally good practice to taper the medication before discontinuing.

Are Cholinesterase Inhibitors and Memantine Effective for Disruptive Behaviors?
The drugs are often advocated for treating behavioral symptoms in patients with AD. However the bases for the claims are post-hoc analyses of individual items on behavior rating scales from the sponsored clinical trials. An argument is made that enhanced attention or cognition due the drugs' effects may improve behavior, and this might be true with patients who have delirium.

In one summary analysis, 14 cholinesterase inhibitor trials were identified that assessed effects on behavior in post-hoc analyses, but only 3 showed statistically significant but modest results at improving behavior [65]; none of these effects was large [16]. Another meta-analysis observed a similar, trivial effect in the more mild patients but no effect on behavior in the more severely impaired [66]. The fact that patients were not chosen for having behavior problems and their symptom ratings were rather low at baseline limits reasonable inferences about efficacy. In the one cholinesterase inhibitor trial in which patients were enrolled because of clinically important agitation at a magnitude for which an antipsychotic is indicated, donepezil did not improve behavior over placebo which itself had a moderate effect on improving behavior [67]. Notably cognitive function was improved with donepezil in this trial suggesting that the effects on cognitive do not influence behavior. Another randomized clinical trial similar in design to this one did not demonstrate that memantine was effective at improving behavior either [C. Fox et al., unpubl. data]. As many patients would already be treated with cholinesterase inhibitors and memantine, the opportunities to actually use the drugs specifically to treat agitation and disruptive behavior are limited. Indeed, a clinician might want to be sure that the cholinesterase inhibitor is not exacerbating the agitation.

Current Pharmacological Treatment Recommendations

When definitive treatments are lacking, treatments with modest effects and benefits can be seen as important for patients, and especially when the treatments appear safe or adverse events are seen as manageable. Within this context, and considering the absence of alternatives pharmacological treatments, donepezil, rivastigmine, galantamine and memantine can be considered drugs of choice for AD.

There is little objective reason to choose one cholinesterase over another for mild to moderate AD, or for severe AD to a priori choose memantine over donepezil. The three cholinesterase inhibitors are more similar than different from each other with regard to efficacy and adverse effects.

Donepezil, however, dominates by far the North American market, and this may be due largely to its earlier marketing, once-per-day dosing and option to adjust the dose only once. Nevertheless, the variability in the effectiveness of the drugs for individual patients can be substantial, and what appears to work for one person may not work for another. Consequently, treatments must be individualized. It is inappropriate, however, to combine two cholinesterase inhibitors based on current considerations of their different mechanisms of action.

In principle, either memantine or donepezil is more effective than placebo in severe AD. The effects appear roughly equivalent although with differences in tolerability that may favor memantine, and possibly differences in cognitive effects that might favor donepezil. It is notable that head-to-head trials comparing the two drugs have not been performed. Similarly, evidence is lacking that the two drugs together are more effective than either alone because the necessary combination studies have not been performed wherein the combination is compared to each drug alone and to placebo. The one 'add-on' trial that randomly assigned memantine or placebo to patients already maintained on cholinesterase inhibitors (for an average of over 2 years) did not address this.

In practice in the USA, however, it is most common for memantine to be added to donepezil, or, alternatively, it is very uncommon to see memantine prescribed alone. Cholinesterase inhibitors may have been added first, perhaps when the patient was within the mild to moderate stage, and memantine then added when the patient transitions into a more severe stage. It is particularly common however for a physician in the USA to add memantine even earlier for patients with mild AD as well.

Treatments for Disruptive Behaviors

A majority of patients with AD develop significant behavior disorders over the course of their illnesses. These include disturbed perception, thought content, mood, and behavior. Disruptive behaviors may be grouped into those that are mainly psychologically expressed such as depression, anxiety, persecutory ideation, auditory and visual hallucinations and apathy, and those that are observable such as agitation, wandering, restlessness, aggression, screaming. These behaviors may be chronic, acute, waxing and waning, and may be particularly difficult to treat. Both acute and chronic agitation is probably most common; aggression and delusions are most disturbing.

In the USA no medication is approved for treating behavioral problems in AD. One antipsychotic, risperidone, is specifically approved in several countries, e.g. UK, Spain, and Canada, for the treatment of agitation, psychosis, or behavioral and psychological symptoms of dementia (often referred to as BPSD) occurring in AD. Thus a range of medications including antipsychotics, anticonvulsants, antidepressants, benzodiazepines, antihistamines, and, as discussed above, cholinesterase inhibitors and memantine have been advocated by experts at various times.

Antipsychotics

The demonstrated effects of both conventional and atypical antipsychotics for treating primarily agitation and delusions are modest at best. Trials that show efficacy have been mostly in nursing home samples and generally involve a few points of improvement on a behavior rating scale. Average effects with risperidone and aripiprazole are slight at best when assessed by meta-analyses, and there is little consistent evidence for olanzapine and quetiapine.

Adverse Events and Safety Warnings with Antipsychotics

Moreover, antipsychotic medications are associated with increased mortality and cerebrovascular adverse events or stroke when administered to elderly patients with dementia or AD. The FDA-approved prescribing information for both atypical and conventional antipsychotic medications contain a warning for mortality in bold print and boxed. The risk of death is increased from approximately 2.6% to approximately 4.5% during an average of 10 weeks of therapy [68]. Long-term observations suggest that continuing therapy is associated with a continuing increased risk for mortality [69, 70]. In epidemiological and surveillance studies, haloperidol appears to carry greater risks than risperidone or other atypical antipsychotics with respect to mortality. Other adverse events include oversedation, syncope, metabolic syndrome including weight gain, hypertension, neuroleptic malignant syndrome (thermal dysregulation), extrapyramidal movement disorders (e.g. dystonia, rigidity, gait disorders) other cardiovascular effects, pneumonia, and urinary tract infections. Particular adverse effects are more or less associated with particular drugs.

Use of antipsychotics to treat agitation, aggression and psychosis in patients with AD is clearly undesirable. Yet alternatives in terms of medications are limited. Sometimes it may be necessary, and use should be limited to patients for whom psychosocial, behavioral, or environmental interventions have failed. When needed, treatment should be with lowest doses for shortest periods, and the need for continuation frequently reassessed.

Antidepressants

At least 25% of many dementia patient samples are maintained on antidepressants; more than are treated with antipsychotics [71]. They have been advocated for treating not only depression, but depressive symptoms, apathy, agitation, and aggression. Clinical trials, however, have been largely negative, demonstrating no benefits for treatment of depression in AD with antidepressants [72–74]. This has been most recently confirmed in a trial in which neither sertraline nor mirtazepine was more effective than placebo [75], although mirtazepine may have had a sedative effect. There is no consistent evidence base for the use of antidepressants for treating either depression or agitation in AD. In a small trial of hospitalized dementia patients with aggression and psychosis, citalopram was equivalent to perphenazine and significantly better than placebo at improving agitation or aggression [76].

Trazodone is an antidepressant that is often used as a hypnotic or for agitation or anxiety in patients with dementia. Compared to placebo there were no benefits on neuropsychiatric symptoms in two trials [77]. Two trials that compared trazodone to haloperidol found no significant relative benefit or harm between these two treatments [78].

Anticonvulsants
A uniquely designed recent trial that assessed divalproex to prevent the onset of behavior problems over 2 years led to no effect, worse cognition and smaller brain volumes. The anticonvulsant divalproex in particular cannot be recommended. Others are unknown.

Summary Recommendations – Treating Disruptive Behaviors
The pharmacological treatment of delusions, aggression, agitation, and depression in people with AD is difficult. Non-pharmacological, psychosocial, environmental, and psychotherapeutic interventions may be preferable to medications and can be effective. Judicious use of psychotropic drugs might be helpful for some patients. However, treatment effects should be frequently and carefully assessed to determine that the patient is benefiting and not having impairing adverse effects. Medications that are not effective should not be continued. Patients that appear to be benefiting should be assessed frequently to assess need to continue or to try to withdraw.

Part 2: Prevention Trials

Prevention trials have a unique place in AD drug development. There is an obvious urgency to be able to prevent or delay the onset of mild cognitive impairment or dementia. It seems desirable to intervene during the illness long pathological but still preclinical phase, when disrupting pathogenesis may be more effective than perhaps it would be further along in the disease process and when symptoms are evident. Prevention trials, however, need to enroll large numbers of participants without cognitive impairment and follow them for a number of years in order to assess the drugs' potential to delay the onset of cognitive impairment. The long-term safety of the test drug, therefore, becomes particularly important. It is not surprising that drugs used in prevention trials are either marketed drugs for other indications, or they are vitamins and food supplements, i.e. nutraceuticals. It is unlikely that a phase 2 or 3 drug in development could be considered safe enough for use in a prevention trial. Thus the drugs used in recent prevention trials are conjugated equine estrogens, the marketed non-steroidal anti-inflammatories, naproxen and celecoxib, the cholesterol-lowering simvastatin, and vitamins (table 3).

The trials enrolled from 1,000 to 20,500 participants, treated and followed them from 4 to 9 years. In some the AD prevention study was nested as a component of

Table 3. AD prevention studies [adapted with permission from 89]

Study	Inclusion criteria	Age years	Sample size	Length	Outcomes	Status
ADAPT/naproxen, celecoxib [119]	First-degree relative with AD	≥70	2,528	5–7	AD, cognitive decline	Early termination
GEM/ *G. biloba* [47, 120]	Asymptomatic 60%, MCI 40%	≥75	3,072	5	AD, cognitive decline, cardiovascular	No significant effects
GUIDAGE/*G. biloba* [121]	Memory complaints	>70	2,854	4	AD	No significant effects
Physicians Health Study-II/vitamin E, folate, β-carotene [122]	Asymptomatic	>65	10,000	9	Telephone cognitive testing	Ongoing
Heart Protection Study/vitamin E, C, β-carotene, simvastatin [123, 124]	Asymptomatic with cardiovascular risk factors	40–80	20,536	5	AD, telephone interview for cognitive status (TICS)	No differences
PreADVISE/selenium, vitamin E [125]	Asymptomatic, males only	≥60	10,400	9–12	Dementia onset, cognitive tests	Terminated
HERS/estrogen medroxyprogesterone (MPA) [126]	Asymptomatic, females	mean = 67	1,060	4.2	Cognitive tests	Improvement on one test
WHI-MS/estrogen + MPA	Asymptomatic, female	65–80	4,532	4–5	AD and MCI, cognitive scores (add on)	Increased risk for MCI/AD, worse scores with HRT
WHI-MS/estrogen alone [127]	Asymptomatic, female	65–80	2,497	4–5	AD and MCI, cognitive scores (add on)	Increased risk for MCI/AD, worse scores with HRT

an even larger trial. The main outcomes in all trials have been onset of dementia or mild cognitive impairment and thus far showed null results. Except for one, GuidAge which was funded by a pharmaceutical company, the trials have been government-funded with the medication usually supplied by the manufacturer.

Even though the drugs were marketed and considered safe for these trials, it is notable that the trials with estrogens and NSAIDs were stopped early because of safety concerns, and that the former trials were associated with a worsening of cognition as well. Thus, the presumption of safety for AD prevention with marketed drugs did not work out.

In sum, prevention trials with marketed drugs might be considered long shots done with the hope of being successful enough that something is learned. In retrospect, these trials represented attempts to jumpstart this area of therapeutics. The sentiment that prevention trials are an imperative is prevalent.

Prevention Trials with Ginkgo biloba

Perhaps because of its popularity and perceived safety there have been three prevention trials undertaken using EGb 761 at doses of 240 mg/day. A trial involving 118 participants without MCI or dementia, all older than 85 years, randomized to receive *G. biloba* extract or placebo and followed up for 42 months, showed a non-significant effect for *G. biloba* to delay progression to mild cognitive impairment [49]. Of potential concern, however, was that more ischemic strokes and transient ischemic episodes occurred in the *G. biloba* group. The GEM trial randomized 3,069 persons to *G. biloba* extract, EGb 761, or placebo who had no cognitive impairment or MCI for a median duration of more than 6 years and found no clinical effects for the extract on cognition or time to dementia [47]. A second prevention trial, GuidAge, conducted in France and involving 2,854 participants with memory complaints or mild cognitive impairment randomized to *G. biloba* or placebo and followed for over 5 years [48], also failed to find effects for ginkgo on the primary outcome of time to onset of AD or other dementia [Ipsen Press Release, June 22, 2010].

Part 3: Current Treatment Development

The development of pharmacological treatments for AD is driven by the enormous need, a large market, and high profit potential. Thus it is a major academic and industry effort. Rather frustratingly, despite major advances in the science of Alzheimer pathogenesis, there have been no practical treatments achieved in over 20 years of effort. Notably, a new drug has not been marketed since 2003. As of 2008, two reviews on drug development identified over 100 [79] and 172 [2] drug development failures in AD. Moreover, drug discovery in general – not just neuroscience discovery – is difficult, complicated, takes from 10 to 15 years from discovery to development, and has a failure rate over 90%, mostly due to lack of safety and efficacy [80].

The barriers to developing a successful drug for AD in particular are substantial. They range from lack of clinical definition of AD, its pleomorphic presentation consisting of several clinical phenotypes that can be targeted, and far too many potential biological targets, none of which have been validated.

Regulatory Requirements

Regulatory criteria for marketing of symptomatic and disease-modifying therapies for AD require demonstrating cognitive efficacy and improvements in function (i.e. activities of daily living, and often evidence of overall clinical improvement or less overall decline), accompanied by adequate evidence of safety [1]. The criteria issued by FDA as draft guidelines and by the EMA as formal guidelines have tended to both facilitate drug development by providing a roadmap of sorts, and constrain development programs by limiting how drugs with vastly different actions for a pleomorphic disorder can be developed. Effectively, the guidelines encourage very similar programs for very different drugs. See 'Guideline on Medicinal Products for the Treatment of Alzheimer's Disease and Other Dementias', CPMP/EWP/553/95 Rev. 1 http://www.emea.europa.eu/docs/en_GB/document_library/Scientific_guideline/2009/09/WC500003562.pdf.

Pharmaceutical companies and academics, therefore, plan fairly standard protocols. For example, mild to moderate AD is indexed by MMSE scores from about 10 or 14 to about 24 or 26. Whether the efficacy of a drug is to be tested over the short or long term, essentially the same standard outcomes are used: the Alzheimer's Disease Assessment Scale – Cognitive Portion (ADAS-cog) [81]; one of two ADL scales, i.e. the Alzheimer's Disease Cooperative Study Activities of Daily Living (ADCS-ADL) scale or Disability Assessment for Dementia (DAD) [82]; a clinician's global assessment (known as a Clinician Interview Based Impression of Change with Caregiver Input (CIBIC+), or ADCS Clinical Global Impression of Change (ADCS-CGIC)) [83], or the Clinical Dementia Rating (CDR), an interview-based overall dementia severity assessment [84].

Phase 2b and 3 trials are commonly of 6–12 months' duration for symptomatic drugs, and 18 months for drugs that could be considered 'disease-modifying', although trial length does not determine the intention for a disease-modifying trial. Two positive, pivotal phase 3 studies, or one phase 3 and one phase 2b trial, are typically required for approval. An open-label follow-up period could be considered as well for further safety assessments. More recent development programs have tended to include more mild AD patients, interpreted as MMSE ≥20, amnestic MCI, MCI due to AD, prodromal or early AD (some supported by a positive putative biomarkers, specifically cerebrospinal fluid Aβ, tau, or an MRI volumetric parameter [1].

For EMA, cholinesterase inhibitors are considered standard care (despite their controversial status) such that a drug being developed needs to be assessed also in a phase 3 add-on design wherein the new drug is used in a placebo-controlled trial in patients already maintained on cholinesterase inhibitors. This occurs de facto, however, as it is very difficult to recruit to long-term clinical trials AD patients who are not taking these drugs.

Another difference between the EMA and FDA guidelines is that EMA suggests a method for disease modification and FDA does not. Under their view, a biomarker could be defined that can be correlated with the presumed mechanism of action of the drug (e.g. CSF Aβ, amyloid PET imaging or hippocampal volumes), and ultimately, it is hoped, with the clinical efficacy of the drug. So far, however, no such relationship has been observed in other than small subgroups of larger trials that showed null, or negative, results on the primary outcomes for the drug in question. For example, an Aβ vaccine, AN1792, reduced plaque density [85]; a trial including only 29 patients showed a small decrease in fibrillar Aβ on PET imaging with bapineuzumab after 18 months of treatment [86]; there was a potential trend depending on the statistical analysis for tramiprosate to be associated with increased hippocampal volume in a subgroup [87]. All these observations, however, were unrelated to clinical improvement with the drug (which did not occur) and thus does not support even a hypothesis for the biomarker in question, let alone a validated biomarker. Regardless, biomarker or not, marketing approval can only be obtained if the cognitive symptoms of AD are significantly improved compared to placebo.

One method for validating a biomarker for a particular drug development program is to first observe the change in biomarker in phase 1, validate it as above in the phase 2 proof of concept studies, and then use it in phase 3. This would require, however, that the drug show efficacy in phase 2, something that has not been accomplished in unbiased, randomized clinical trials for other than cholinesterase inhibitors.

'Typical' Development Programs

A typical development program for a putative, disease-modifying new AD therapy might consist of the following to comply with EMA and FDA requirements. The phase 1 program would include single-dose, single-escalating dose, and multiple-dose studies in healthy volunteers or in patients with mild AD who have the capacity to provide informed consent. These studies might include 24–50 participants. Potentially, studies in renal and hepatic impairment, and food interaction studies are performed along with various drug-interaction studies (e.g. with antihypertensives, digoxin, warfarin, antidiabetics, antipsychotics, antidepressants).

Phase 2

Phase 2 studies are generally multicenter, double-blind, placebo-controlled, parallel-group trials with an AD target population, and further defined by level of severity using the MMSE (e.g. generally, mild to moderate AD, but sometimes restricted to mild or severe). The phase 2a trials may have durations of 3–6 months and are done mainly for safety, tolerability, dose finding, and as 'proof of concept' or initial demonstration of efficacy. They might typically be dose-ranging trials done with two or more doses of the investigational drug compared to placebo. An open-label follow-up period of up to 12 months could be used to gain longer term, although uncontrolled, patient exposure to the drug for safety purposes. Two to four trials could be undertaken in phase 2 to help round out safety and tolerability and to better determine a preferred dose for phase 3.

The efficacy measures used in phase 2 are generally the same as those that are used in phase 3. A conundrum is that in most AD development programs, efficacy is not seen in phase 2 but the sponsor progresses to phase 3 nonetheless, usually justifying the decision based on safety, post-hoc subgroup analyses for efficacy, a finding with a putative biomarker in a subset, or on theory. Indeed, it is common to see drugs progressed to phase 3 without adequate knowledge of dosing or evidence for efficacy.

Lately, the target populations for phase 2 have been expanded from mild to moderate AD to mild AD, prodromal AD, or mild cognitive impairment due to AD. These newer terms refer mainly to AD before dementia onset and are characterized in common by progressive episodic memory impairment with only mild impairments in executive function or instrumental daily activities. These diagnoses can be further restricted by requiring patients entering the trials to have a positive putative biomarker such as low Aβ and/or high tau in the CSF, or significant ligand retention on Aβ PET imaging. The length of phase 2 trials for some drugs has been extended to 18 months. Thus, a current phase 2 profile for AD drug development is emerging that looks very similar to phase 3 except for a smaller sample size.

Phase 3

The two positive phase 3 trials that are required for approval have to be double-blind, placebo-controlled, parallel-group trials. They will of necessity be performed in multiple sites across several countries, although the FDA requires a substantial proportion of patients in a drug development program to be treated in the USA. An open-label follow-up period of about 1 year can be considered for safety purposes, but thus far very few phase 3, 18-month trials have been extended that far into open label, perhaps because none has shown significant outcomes in favor of the drug. As above, these trials can be 'add-on' trials in that patients already being treated with cholinesterase inhibitors can be randomized to the experimental drug or placebo.

Reflecting phase 2, two primary outcome measures should be a cognitive measure and either a functional or global assessment. From EMA's perspective, for a disease-modifying drug a – presumably validated – biomarker should be identified that is related to action of the drug. For a disease-modification claim the drug would show significant outcomes on both the clinical and the biomarker outcomes. It remains to be seen however if this scenario is accomplished, whether or not the regulatory authority will indeed judge the drug as 'disease-modifying'. For example, it is possible to envision an Aβ antibody further lowering CSF Aβ or decreasing brain Aβ by PET imaging while improving cognition and still not receive a 'disease-modifying' claim precisely because the relevance of the biomarker as an index of pathological progression or on clinical action could be questioned. EMA, however, might allow a claim of 'delay in disability' instead. These two claims, disease modification and delay in disability, are of marked commercial importance as pharmaceutical companies believe they will improve sales when compared to a simple claim such as 'for the treatment of AD'.

Safety and Risk Management

In general, safety assessments for AD trials are standard as per ICH or FDA general safety guidelines, but depending on the mechanism of action or known safety profile of the investigational compound, certain adverse effects may be proactively sought through prospective screening. Recently, the FDA required that in trials for most anti-Aβ agents MRIs be performed every 3 months in order to detect cases of amyloid-related imaging abnormalities, including vasogenic edema, microhemorrhages, and hemosiderin deposits. Thus, for example, vasogenic edema was detected in an early phase 2 study of the Aβ antibody gantenerumab leading to the discontinuation of the higher dose [88]. As the elderly population is also a vulnerable one, it is important to carefully monitor for cardiovascular and neurological adverse effects. For many central nervous system drugs, either the sponsor or the FDA might require more extensive clinical cardiac monitoring or monitoring for metabolic changes.

Overview of Alzheimer's Disease Drugs in Development

The uncertainty about the causes and pathological progression of AD, the realization that the illness is clinically and biologically pleomorphic, means that there are many potential molecular targets for therapy. In a milieu in which there are many targets it is difficult to discover relevant and validated clinical targets. Therefore, the current status is that there are many kinds of drugs under development. They can be grouped for convenience into approaches directed at amyloid-β protein, tau, neurotrophic and neuroprotective, and other small molecules, many directed toward various neuroreceptors. Some potential approaches and drugs are listed (table 4).

The Amyloid Hypothesis Dominates Drug Development

In the midst of uncertainty about the causes of AD, the amyloid hypothesis dominates drug development and is the most researched conceptual framework for the disease [89]. Consequently, preclinical and translational drug development for AD is amyloid-β-focused. Some scientists might argue that AD drug development is too much amyloid-β centric.

The core hypothesis is that the biological and clinical expression of AD results from the accumulation of Aβ in the brain due to increased production or impaired clearance of Aβ peptides generated by variations in the processing of APP, a neuronal membrane protein, and is the primary influence driving pathogenesis [89, 90]. Although $A\beta_{40}$ is the most prevalent form of Aβ peptide, $A\beta_{42}$ and perhaps other forms have a greater propensity to aggregate into toxic oligomers and fibrils and to form amyloid deposits. Further expression of the amyloid hypothesis involves the type of Aβ that may explain or cause the illness: amyloid plaques, protofibrils, increased $A\beta_{42}$, the relative amounts of $A\beta_{42}$ in relationship to $A\beta_{40}$, and oligomeric forms of Aβ. In this context, for example, Aβ oligomers show direct synaptic toxicity effects and plaque-derived Aβ fibrils are proinflammatory and neuron toxic.

Another take on clearance of Aβ is that the most important genetic risk factor, the ApoE ε4 allele, is associated with increased amyloid burden in brain, perhaps because its product, apolipoprotein E4, is less able to clear Aβ from brain than apolipoprotein

Table 4. Recent and current anti-amyloid approaches

Method	Drugs in development (examples)
Active and passive immunization	
Aβ antibodies by active immunization, i.e. vaccine (all are in phase 1 or 2 of development)	Aβ vaccine, AN1792 (Elan/Wyeth) – unsuccessful in phase 2 CAD-106 (Novartis/Cytos) – in phase 2 ACC-001 (Janssen Alzheimer Initiative) – in phase 2 ACI-24 (AC Immune) AD02 (Affitope) (Affiris/GSK) – in phase 2
Aβ antibodies by passive immunization	Bapineuzumab (AAB-001) (Janssen Alzheimer Initiative) Solanezumab (LY2062430) (Eli Lilly) IVIg (Gammagard) (Baxter) Crenezumab (MABT5102A) (Genentech/AC Immune) Gantenerumab (R1450) (Roche/Morphosys) Ponezumab PF04360365 (Pfizer) GSK933776 (GSK)
Modulation of Aβ production	
γ-Secretase modulators	Tarenflurbil (Myriad) – unsuccessful in phase 3 E2012, E2212 (Eisai) CHF 5074 (Chiesi)
γ-Secretase inhibitors	Semagacestat (LY 450139) (Eli Lilly) – unsuccessful in phase 3 Avagacestat (BMS 708163) (BMS) – in phase 2 EVP-0962 (EnVivo)
BACE-1 inhibitors	HPP854 (TransTech) MK-8931 (Merck) LY 2811376 (Lilly) – discontinued because of preclinical safety CTS-21166 (CoMentis/Astellas)
Inhibition of Aβ aggregation or prevent oligomerization	Tramiprosate (homotaurine) – unsuccessfully tested Scyllo-inositol (ELND005/AZD103)(Elan) – negative phase 2 trial Curcumin Melatonin Resveratrol Thioflavin T PBT-2 (Prana) – metal attenuation of proteins
Enhancement of Aβ degradation	Neprilysin activation Insulin-degrading enzyme (IDE) activation
Other approaches	Atorvastatin (Pfizer) – unsuccessful in phase 2/3 Simvastatin – unsuccessful in phase 3 Rosiglitazone (Avandia, GSK) – unsuccessful in phase 3 Pioglitazone (Takeda, Zinfandel) – planned prevention trial

E3 or E2. Apolipoprotein E4 may be also less stable and efficient at clearing Aβ, and its peptide fragments might be toxic to mitochondria function [91]. Finally, the mutations that cause familial early-onset AD, principally, PSEN 1 mutations, by altering APP metabolism, could increase the production of $Aβ_{42}$ that is more likely to aggregate than $Aβ_{40}$. As is evident, there are numerous potential 'amyloid targets' for drug development. Unfortunately, none has been validated and several drugs shown to be active at the intended amyloid target have not had positive clinical effects.

There are now several clinical examples that suggest that reducing Aβ in the brain, i.e. decreasing production of Aβ in the central nervous system or reducing plaques, first, is not necessarily negative, but, second, that it is also not necessarily associated with improvement [85, 86, 88, 92].

Various 'Anti-Amyloid' Approaches

The basic approaches involve several Aβ-targeted therapeutic strategies, including modulation of Aβ production, inhibition of Aβ aggregation, enhancement of Aβ degradation, and the administration of or raising of antibodies that target Aβ by passive or active immunization. Table 4 lists a range of these approaches to reducing Aβ and some specific drugs either currently in development or recently discontinued. An additional approach to modifying Aβ clearance is through regulating clearance of Aβ from the brain as apolipoprotein E4 may clear Aβ more slowly than the other isoforms [93].

Antibodies against Aβ

Antibodies against Aβ can either be induced endogenously with Aβ vaccines or produced as monoclonal humanized antibodies. In addition, natural immunoglobulins harvested from blood donations contain antibodies to Aβ. The antibodies might bind to and help remove soluble and aggregated Aβ and inhibit generation of toxic Aβ aggregates. The several ways in which Aβ antibodies could mediate amyloid clearance include: (1) antibodies could interact with brain amyloid deposits and initiate phagocytosis by microglia; (2) they could interact with brain amyloid deposits and directly neutralize any toxic effects of the deposits; (3) they could interact with soluble Aβ in the blood, without penetrating brain, and create a 'peripheral sink' effect, pulling soluble Aβ from the brain to the periphery and then clearing it, and (4) antibodies can bind to Aβ oligomers and directly inactivate any toxic effects.

Depending on their relevant clinical action and one's point of view, Aβ antibody therapy might be viewed as symptomatic or disease-modifying treatments. For example, binding to oligomers may have no impact on Aβ production but an effective symptomatic effect as it might prevent Aβ from binding to neuronal membranes causing toxicity.

There are advantages and disadvantages to each immunization strategy with important practical implications. Active vaccine immunotherapy can result in constant high antibody concentrations, requiring few follow-up visits. Passive immunotherapy can more easily target specific Aβ epitopes with specific antibodies and can provide direct control of antibody concentrations. It is difficult to rapidly reduce antibody concentrations and limit adverse effects due to the infusions. Vaccines also require that the patient is able to generate an appropriate response. As the elderly have reduced responsiveness to vaccines, this might be limiting to the efficacy of vaccines. The administration of antibodies is inconvenient as it requires intravenous infusions by a nurse either at home or in the clinic every 4 to 13 weeks and most likely more frequent subcutaneous infusions perhaps every 2–4 weeks. Immunological reactions are a potential problem and will likely vary with the antibody or vaccine used.

Some Selected Aβ Vaccine Approaches

Aβ vaccines are in general earlier in development than antibodies. The notable failure of AN1792 (Elan, Inc.), a full-length $Aβ_{42}$ vaccine, substantially informed subsequent vaccine development. In a phase 2 trial that was stopped early, there were conflicting trends for slower cognitive decline on one cognitive outcome, the Neuropsychological Test Battery (NTB) [94], but opposite or non-significant effects on the ADAS-cog and other outcomes in the vaccinated patients compared to placebo-treated patients [92]. However, 6% of patients developed meningoencephalitis that was possibly due to a T-cell-mediated response and microglia activation against the synthetic $Aβ_{42}$ vaccine [95], thus halting further development.

Newer vaccines are designed with AN1792 in mind. For example, ACI-24 (AC Immune) is composed of a palmitoylated peptide corresponding to the first 15 amino acids of Aβ peptide and attached to liposomes. The rationale is that the 1–15 amino acid sequence contains the B-cell epitope, is not a site reactive to T cells, and might therefore avoid meningoencephalitis [96]. Vaccines that are in phase 2 trials include CAD-106 (Novartis and Cytos), ACC-001 (JAI), and AD02 (Affiris).

Some Selected Aβ Antibody Approaches

Some selected Aβ antibodies in phase 2 and 3 of development are listed in table 4. The Aβ antibodies differ from each other in part by their epitopes on Aβ fragments, their potency, and their pharmacokinetics. For example, bapineuzumab targets the N-terminus and amino acids sequence 1–5; solanezumab targets the mid-domain amino acids sequence 16–23; ponezumab may target the C-terminus amino acids 33–40, and crenezumab may target both the N-terminus and central domain. Examples of other differences among antibodies are that bapineuzumab binds to amyloid plaques, solanezumab is said to preferentially bind to soluble Aβ [97], and crenezumab is considered to bind equally well to Aβ monomers, oligomers, and may inhibit aggregation into fibrils.

Bapineuzumab [98] and solanezumab are in phase 3 trials. The basis for entering phase 3 with solanezumab was a 12-week phase 2 trial with 52 patients that showed a lowering of Aβ without clinical efficacy and presumably adequate safety. The basis for the large phase 3 program with bapineuzumab was mainly a phase 2 multiple ascending dose trial involving four panels of approximately 60 patients randomized to drug or placebo and followed for 18 months [98]. None of the doses showed efficacy; the higher doses showed adverse effects including vasogenic edema, seemingly mainly in ApoE ε4 allele carriers. The phase 3 program was planned as two 18-month trials, comprising ApoE ε4 carriers, and two larger trials of 1,200 patients each comprising only non-carriers in which higher doses were used. The higher dose in the latter trial had to be discontinued because of vasogenic edema. The trials are continuing and will complete by mid-2012.

Crenezumab (MABT5102A, Genentech) is in a larger phase 2 trial involving 372 patients randomized to one of two doses of crenezumab or placebo and treated for 16 months; and in a separate biomarkers study involving 72 patients. The former includes volumetric MRIs with all patients; the latter requires CSF, FDG-PET, and vMRIs with all patients.

In a phase 1 multiple ascending dose study, the higher doses of gantenerumab were associated with vasogenic edema after about four monthly infusions [88]. Gantenerumab is in a phase 2 trial of prodromal AD with Aβ-PET imaging and CSF Aβ and tau measures in addition to cognitive outcomes.

Intravenous immunoglobulin (Gammagard, Baxter), a polyclonal antibody, binds to Aβ and prevents aggregation. In humans, CSF Aβ levels decreased and plasma levels increased with IVig administration over 6–18 months which was interpreted as the clearance of Aβ from brain, and cognitive improvement was reported at 6 months in a phase 2 trial involving 24 patients [99]. There is an ongoing phase 3 trial of about 360 mild to moderate AD patients in which doses of 200 and 400 mg/kg are administered every other week.

The results of four bapineuzumab and two solanezumab phase 3 trials will provide information on the viability of passive Aβ antibody approaches. None of the antibodies has been associated with efficacy yet.

Secretase Modulators and Inhibitors

As Aβ fragments are created by the action of secretases on the large APP, inhibition or modulation of the actions of the enzymes could inhibit or alter Aβ production, including the important $A\beta_{42}$ peptide, and this has been a major treatment development focus. The involved proteases are γ- and β-secretase (i.e. β-site APP cleaving enzyme 1 or BACE 1). A third protease, α-secretase, competes with β-secretase for APP substrate and can preclude the production of Aβ by cleaving the peptide in two. Thus a major research effort is aimed at reducing $A\beta_{42}$ production with γ-secretase inhibitors and modulators and with BACE-1 inhibitors.

γ-Secretase Modulators and Inhibitors

γ-Secretase inhibitors decrease amounts of the presumed toxic $A\beta_{42}$ but they also inhibit Notch, an important signaling molecule. Newer γ-secretase allosteric modulators may exert their effects exclusively on the γ-secretase complex and may spare Notch signaling avoiding this problem.

Tarenflurbil

Tarenflurbil (R-flurbiprofen), the racemate of which is a marketed anti-inflammatory agent, is an allosteric modulator of γ-secretase. The L-isomer possesses most of the anti-inflammatory properties. Phase 1 studies indicated that tarenflurbil could be measured in CSF but that it was not clear that CSF Aβ was lowered. A phase 2 trial in which 210 patients were randomized to higher or lower doses or placebo and followed for 12 months indicated no overall effects [100]. When the sample was split into mild and moderate AD, the mild patients at the high dose showed effects on activities of daily living and global assessment, but not on cognition (the moderate patients, however, showed effects in favor of placebo). A phase 3 program focused on mild AD patients – i.e. MMSE from 20 to 26 – and the higher 800-mg b.i.d. dose. It included one 18-month long trial of 1,684 patients and another planned for about 900. Although the placebo group deteriorated as expected, the outcomes were wholly negative [101]. Other γ-secretase modulators are under development.

Semagacestat

At least two γ-secretase inhibitors have entered phase 3 clinical trials including semagacestat and BMS-708163. In one phase 3 program of the γ-secretase inhibitor semagacestat was ended due to cognitive toxicity – i.e. actually worsening cognition – and perhaps increased rates of squamous or basal cell skin cancer that might have occurred in 5%. In one of the 18-month long trials, involving 1,536 patients randomized to receive 140 or 100 mg/day of semagacestat or placebo, the placebo group worsened by an expected 6.2 points on the ADAS-cog scale but both drug groups worsened more, by 7.7 and 7.3 points for patients receiving the 140- and 100-mg/day doses respectively. Semagacestat seemed to do what was expected of it by reducing blood and CSF Aβ, suggesting reduction in Aβ synthesis, but was not associated with reducing fibrillar Aβ as evidenced by lack of changes with amyloid PET imaging. It was associated with reduced glucose uptake on FDG-PET in brain regions associated with AD.

During a 32-week follow-up period after the trial was stopped, patients who had been treated with the drug did not improve cognition toward baseline or catch up to the placebo group, i.e. there was no reversal in worsening. The negative or absent effects on cognition could be due to other effects of blocking γ-secretase, unrelated to its actions on APP as γ-secretase may regulate the activity or metabolism of other proteins such as Notch, or many other substrates.

BMS-708163

BMS-708163 completed a phase 2 trial very similar in design to semagacestat's phase 3 trials but with a much smaller sample of 209, and with safety rather than efficacy as the primary outcome. In this trial of mild to moderate AD patients – i.e. MMSE 16 to 26 – the two higher doses of 100 and 125 mg daily were not well tolerated, but the lower doses of 25 and 50 mg/day were. There was a trend for cognitive worsening in patients receiving one of the two higher doses, whereas patients receiving the two lower doses showed no significant cognitive changes. None of the patients, regardless of treatment dose, had significant changes in brain volume, but a subset receiving the highest doses showed trends for changes in CSF Aβ. Twice as many patients receiving the higher doses discontinued compared to patients receiving the lower doses or placebo, most commonly due to gastrointestinal and dermatological adverse effects. Some patients developed squamous cell or basal cell cancer. It is questionable, therefore, whether there is evidence of an effective dose range with adequate safety and tolerability, or of expected γ-secretase inhibition or biomarker effects. Although touted as having substantially less activity on Notch, BMS-708163 ultimately may be more similar to semagacestat than originally anticipated.

A second phase 2 trial testing the 50 mg/day dose against placebo in 270 patients with prodromal AD – with MMSE from 24 to 30 – which is essentially 'MCI due to AD' by ADNI criteria with positive CSF biomarkers (either $A\beta_{42}$ levels <200 pg/ml or T-tau/$A\beta_{42}$ ≥0.39), is ongoing.

Toxicity of γ-Secretase Inhibitors

The current failures prompt the question of whether a γ-secretase inhibitor can be safe. Substantial toxicity may be unavoidable with γ-secretase inhibitors as they would most likely inhibit other substrates. One concern is γ-secretase cleavage of the Notch receptor, and consequent impairment of cell differentiation and growth, and metaplasia.

BACE-1 Inhibitors

BACE-1 inhibitors are less developed as therapeutics and as with γ-secretase inhibitors should inhibit the formation of Aβ peptides. The target is perceived as more desirable than γ-secretase because it is a single protease and cleavage of Notch is not an issue. A main barrier to their development is the difficulty in developing molecules small enough to penetrate brain, and then safety concerns. Such molecules are now entering phase 1 and 2 and long-term safety and efficacy remain to be determined. For example, LY2811376 was shown in humans to produce sustained lowering of CSF and blood Aβ, but development was discontinued because of preclinical ocular toxicity [102]. A Merck compound MK-8931 also shows Aβ lowering and is planned for phase 2 in 2012.

In addition to small molecule inhibitors, another approach to BACE 1 inhibition is to develop an antibody that is specific to both BACE 1 and the transferrin receptor that allows the complex to be transferred across the blood-brain barrier. There are many programs involving about nine companies.

Fibrilogenesis or Aβ Aggregation Inhibitors

As discussed above, $A\beta_{42}$ can form soluble oligomers that are thought to be differentially neurotoxic from that of large fibrils from plaques and may cause synaptic dysfunction. Fibrilogenesis inhibitors purportedly interfere with the Aβ peptide interactions. In theory this would inhibit the production of toxic oligomers and the evolution of amyloid plaques. Some putative inhibitors of Aβ aggregation or fibrilogenesis include: tramiprosate (3-amino-1-propanesulfonic acid, homotaurine), scyllo-inositol, curcumin, melatonin, and resveratrol. The latter three, however, have other actions and have not yet been systematically studied.

Tramiprosate

Tramiprosate, the amino acid analog, homotaurine, was reported to bind to Aβ monomers preventing oligomeric fibrils from developing. In a placebo-controlled, 12-week dose-ranging, phase 2 trial involving 58 patients receiving one of three doses of tramiprosate or placebo, there were no clinical effects but CSF $A\beta_{42}$ was reported decreased in the moderately impaired subset of patients [103, 104], although the meaning of this is unclear. In two phase 3 trials, however, each including about 1,000 patients, it did not demonstrate efficacy [87].

Scyllo-Inositol

Scyllo-inositol (ELND005) is a cyclohexanehexol isomer that may stabilize Aβ into non-toxic conformation and inhibit Aβ fibril aggregation [105]. A larger phase 2 trial, however, was wholly negative, after the two higher doses were stopped because of unexpected deaths and infection [106]. Originally, 351 mild to moderate AD patients were randomized to one of three doses (i.e. 250, 1,000, or 2,000 mg/day). The two higher doses were discontinued because of a high rate of infections and 5 and 4 deaths in the two higher doses compared to none and 1 in the placebo and lower dose groups.

The 88 patients receiving 250 mg of the drug and the 83 patients on placebo did not differ on the primary outcomes, the NTB and ADCS-ADL, or secondary outcomes. In an unusually selected subgroup of 46 of 71 mild patients – i.e. MMSE 23–26 – who completed the 18-month protocol, the 24 drug-treated patients showed a better effect on the NTB than the 22 placebo-treated patients but no other effects. Another scyllo-inositol-treated subgroup showed increased brain ventricular volume, and from a group of 33 who had CSF measures at baseline, the 20 who had measures at 18 months showed a non-significant trend for lower CSF $A\beta_{42}$ and no change in tau. Based on the negative results the drug is being taken to phase 3 development (http://newsroom.elan.com/phoenix.zhtml?c=88326&p=irol-newsArticle&ID=1458241&highlight).

PBT-2

Another approach to Aβ modulation is by preventing Aβ from binding to copper or zinc that then causes aggregation to oligomers and fibrils. PBT-2 (Prana) binds to copper. In a 12-week phase 2 trial including 78 patients, there was no significant effect on the main outcomes, but there was improvement on two subtests of the NTB battery and reduced CSF Aβ [107]. Although development plans were announced, the drug has not been advanced further.

Other Anti-Aβ Approaches

Other approaches include antagonists to the receptor advanced glycation end product (RAGE) that are increased in AD and bind amyloid fibrils; RNA interference to block the production of APP; decreasing cholesterol, a range of other aggregation inhibitors; enhancing apolipoprotein function to better clear Aβ, and the use of insulin-degrading enzyme or neprilysin to degrade Aβ. The RAGE inhibitor PF-04494700 did not show efficacy in an 18-month phase 2 trial involving about 400 patients and indeed showed cognitive toxicity. The higher dose needed to be discontinued midway through the trial [Galasko et al., unpubl. data, 2011].

Summary of Anti-Aβ Approaches

Thus far there has not been a successful proof of concept for an anti-Aβ approach and no phase 2 or 3 trial with significant outcomes. Many of the drugs in development have been demonstrated to be biologically active in ways expected from preclinical models. But this effect has not translated into clinical efficacy.

Some developers make an assumption based on the results of phase 2 trials that despite null cognitive or clinical effects, an observed biological effect on Aβ is sufficient to predict success in phase 3. Indeed, none of the ongoing phase 3 programs had positive clinical effects in phase 2 to support them, and some, such as Lilly's, were not intended to be able to detect clinical effects as their phase 2 studies and were limited to about 50 patients in order to mainly be able to demonstrate pharmacokinetic and pharmacodynamic effects on Aβ. Any argument that there was a clinical effect in the phase 2 trials reported is based on post-hoc analyses of subgroups of uncertain validity. Thus, if any of the phase 3 trials are positive, then it will not have been predicted on the basis of phase 2 evidence.

Finally, lowering Aβ in the brain possibly may not be beneficial to patients. With many anti-Aβ drugs there is a fairly prompt reduction in amyloid without obvious cognitive effects.

Tau in Alzheimer's Disease and Anti-Tau Approaches to Therapy

Neurofibrillary tangles are defining histopathological characteristics of AD, and have long been correlated with the severity of dementia [108]. The tangles represent the aggregation and accumulation of hyperphosphorylated forms of the microtubule-associated protein tau. Microtubule-bound soluble tau supports axonal transport. Hyperphosphylation might cause detachment of tau from microtubules which could then lead to the formation of soluble tau aggregates and insoluble paired helical filaments. Disruption of the microtubules and axonal transport, toxic effects of soluble hyperphosphorylated tau and of fibrillar tau could contribute to neurodegeneration. It is not known whether the neurofibrillary tangles are specifically toxic or whether aggregated tau is the pathogenic species in tauopathies, or also a possible mediator of Aβ toxicity in AD. Nevertheless, tau oligomers may be important for neurodegeneration and are potential targets for drug intervention.

A current hypothesis is that Alzheimer pathology starts as pre-tangles in proximal axons of the noradrenergic locus ceruleus, and that pathology spreads by neuron-to-neuron and transsynaptic transport of tau aggregates to the entorhinal cortex, hippocampus, and neocortex [109]. Such a hypothesis, of prion-like cross-neuronal propagation, if true suggests not only several anti-tau protein interventions but also small molecule interventions targeted toward midbrain monoaminergic systems and perhaps with existing neurotransmitter active drugs.

Some 'Anti-Tau' Therapeutic Strategies

As with Aβ, a broad range of therapeutic approaches toward tau can be considered. Those under development include inhibition of tau kinases, enhancement of phosphatase activity to enhance microtubule stability, altering or blockading tau hyperphosphorylation, decreasing or inhibiting tau aggregates and filament formation, enhancing clearance of aggregates with drugs or antibodies. Most of these approaches are in preclinical development although a few have progressed to phase 2 clinical trials [110, 111].

Approaches to suppress tau hyperphosphorylation can involve inhibiting one of several protein kinases, including cyclin-dependent kinase 5 activator 1 (CDK5R1), MAP/microtubule affinity-regulating kinase 1 (MARK1) and glycogen synthase kinase 3 (GSK3), but it is not clear whether inhibiting any one versus several would be sufficient. Aggregation inhibitors could block the formation of soluble tau aggregates and tangles.

Some potential kinase inhibitors include lithium, valproic acid (i.e. divalproex sodium), and methylthioninium chloride (methylene blue). The latter has been reported to dissolve tau filaments isolated from AD brains and to prevent tau aggregation. A phase 2 placebo-controlled trial in 332 subjects with mild to moderate AD showed uncertain results [Wischik, unpubl. data 2008]. Newer formulations of methylthioninium are planned for clinical development. Lithium was demonstrated to be too toxic in a small pilot trial [112]. In another small 12-month long trial, lower plasma levels of lithium were associated with a possible decrease in CSF phosphorylated-tau and questionable cognitive improvement [113]. A long-term trial of divalproex to delay emergence of agitation symptoms and to assess for cognitive outcomes was not only negative but divalproex was associated with decreased brain volumes [114]. There exist marketed kinase inhibitors used for cancer therapy but they may be too toxic for AD.

Davunetide (Allon Therapeutics) is a small peptide analog of neuroactive protein given intranasally or subcutaneously. As with neuroactive protein, it is neuroprotective against Aβ-induced toxicity and protects microtubules against tau-induced damage. A 12-week phase 2 trial was done in 144 amnestic MCI patients wherein davunetide was significantly effective on two memory measures. A small trial with patients with schizophrenia showed a trend for cognitive improvement. Currently the drug is in a 1-year phase 2/3 trial for progressive supranuclear palsy involving 300 patients.

The potential pathological characteristics of extracellular tau and its effects on tangle pathology and neurodegeneration, and the possibility of neuron-to-neuron spread, make tau a potential target for active or passive immunotherapy [115]. Active immunization targeting epitopes of tau oligomers may reduce tau aggregates and slow progression of the tangle-related expression of AD including cognitive impairment.

Tau pathology is not specific to AD however, occurring in other neurodegeneration disorders. Any of these drugs might be useful for any of several tauopathies, including frontotemporal dementia, corticobasilar degeneration, progressive supranuclear palsy and AD, among others. As Aβ and tau pathologies co-occur and are probably synergistic, targeting both together or sequentially may be necessary for efficacy. Thus this area of therapeutics is wide open.

Other Approaches and Small Molecules

As evident above, there are multiple possible targets for AD and include small molecules as well. Most of the small molecules target neurotransmitter receptors. They include nicotinic neuronal receptor agonist, serotonin, 5-HT subtype-active drugs, histamine, H_3 subtype antagonists, metabolic enhancers, vitamins, PDE-4 inhibitors, GABA-A modulators, monoamine oxidase, type B inhibitors (EVT 302, Evotec and Roche), and group 2 metabotropic glutamate inhibitors. Some of these drugs may have direct and indirect effects on Aβ generation or secretase activity. For example, etazolate (EHT 0202) may both allosterically modulate GABA-A receptors and stimulate α-secretase activity. H_3 antagonists may act on H_3 presynaptic autoreceptors to increase cholinergic and monoaminergic neurotransmitter release and are associated with enhanced cognitive function in preclinical models. Several H_3 receptor antagonists are in early stage clinical trials for cognitive improvement in AD. Group 2 metabotropic glutamate inhibitors might combine cognitive effects, neurogenesis effects, and reduction of $A\beta_{42}$.

Nicotinic Neuronal Receptor Agonists

Nicotinic agents currently under development include nicotinic α_7 receptor modulators and nicotinic $\alpha_4\beta_2$ receptor modulators; and may improve attention, arousal, and concentration in adults. Their usefulness in such cognitive impairment syndromes as

AD has not been adequately investigated. Stimulation of the receptors may increase several neurotransmitter including dopamine, serotonin, glycine, glutamate, and GABA. They may affect Aβ processing as well, and have a role in attention-deficit disorder and cognitive impairment associated with schizophrenia. Some examples of α_7 modulators include MEM 3454 (Roche), SSR 180711 (Wyeth), TC-5619 (Targacept), AZ 3480 (AstraZeneca), and EVP-6124 (EnVivo). Some are in phase 2.

5-HT$_6$ Antagonists

Selective 5-HT$_6$ receptor antagonists reverse scopolamine-induced deficits in animal recognition or memory tasks such as the water maze. They may have modulating effects on cholinergic, glutamatergic and monoaminergic neurons. SB742457 is a 5-HT$_6$ antagonist that showed improvement in cognition and global ratings in three phase 2 trials in mild to moderate AD patients [116, 117], although a fourth did not show effects [Maher-Edwards et al., unpubl. data, 2011]. Other 5-HT$_6$ antagonists may be under development with Biotie, Roche, Suven, Wyeth, and Lundbeck.

Challenges of Developing Effective Treatments for Alzheimer's Disease

The considerable challenges to developing effective treatments for AD include the lack of validated drug targets, the pleomorphic clinical targets, and the development of efficient clinical trials designs and outcomes. First and foremost, however, is that we do not know the causes of AD. As discussed above, none of the very broad number of potential drug targets is validated, except for perhaps cholinesterase inhibition with their associated modest efficacy. Establishing validated drug targets requires greater understanding of the illness and the numerous processes leading to illness. For example, the amyloid cascade may be well understood and candidate drug targets are apparent; but it is possible that no intervention in this area will demonstrate efficacy.

The several AD-related clinical diagnoses, including Alzheimer dementia, mild cognitive impairment due to AD, prodromal AD, results in biologically and clinically heterogeneous groups of patients. As examples, patients will show various cognitive profiles, severity of early memory impairments, various genotypes, and various expressions of putative biomarkers. Yet there is no prior knowledge that any particular clinical phenotype will preferentially respond to any given therapeutic approach. Similarly, there is no information that a patient with mild memory impairment will respond differently from one with more moderate impairment, or that one with small hippocampus volumes or low CSF Aβ levels will respond better than a patient with larger volumes or higher levels.

Moreover, the clinical trials have been restricted in design. The designs and outcomes tend to be nearly identical from one program to the next, and are not necessarily relevant to the modeled action of the drugs. For example, nearly all recent phase 2 and 3 trials of drugs that might lower Aβ have been for mild to moderate Alzheimer patients, have an 18-month follow-up, and use the same clinical outcomes. Yet the various drugs act somewhat differently and may have effects at other stages of the illness or in prevention. For the most part the trials designs are not relevant to the known actions of the drugs being tested.

Thus major barriers to successful drug development are our current translational models and translating the way drugs work in the models to humans. It is possible that newer approaches to prevention trials, stratified medicine, and smaller phase 2a trials to gain early signals of potential efficacy may be helpful. The failures in trials have led to many explanations. Although, again, the most likely explanation for the lack of effective medications is that the drugs do not work.

Prospects for Experimental Treatments

Advances in basic and clinical science including imaging and better knowledge and selection of drug targets will strongly drive future drug development. Advances in clinical trials and development methods will be more incremental and based less on science and more on experience. Although it might seem otherwise, many clinical development advances have evolved through the several failures and have improved prospects for identifying effective drugs. Progress will be made in understanding the pleomorphic biological and clinical expression of AD and finding valid treatment targets. When truly effective drugs are tested, however, their effects will overcome the current inefficiencies in clinical development.

Predictions of an effective treatment in the near future can be based only on the current drugs in development. With the exception of some speculative small companies, drug development would not be financed if there were not some preclinical evidence that enthused enough experts to make the investments seem worthwhile. In the end, the advent of an effective treatment for AD may surprise us, coming from outside the current scientific focuses and development programs, and may not have been broadly predicted.

Conclusion

Current pharmacological treatments for AD are the cholinesterase inhibitors and memantine. Alternatives, including vitamins, food supplements and *G. biloba* extract, have not demonstrated an effect either for improving symptoms or delaying onset. The effective pharmacological treatment of behavioral symptoms is a challenge as well. Modest advantages of antipsychotics for delusions or aggression are offset by their considerable toxicity, and they should be used cautiously or avoided. Antidepressants have not been demonstrated effective for depression or agitation in dementia. Anticonvulsants should not be used. There are no demonstrated, effective pharmacological approaches to preventing or delaying onset of mild cognitive impairment or Alzheimer dementia.

The main focus of current drug development is toward the amyloid hypothesis, and has been unsuccessful so far. Various small molecules, active at nicotinic and serotoninergic receptor subsites, are areas of interest. Research into anti-tau approaches is gaining momentum.

Clinical drug development is challenging as there are considerable barriers to finding new drugs in general let alone drugs for AD. Future drug development may be influenced by somewhat different diagnostic criteria than used in the past. Attempts will be made to employ biomarkers for diagnosis, as pharmacodynamic markers, for tracking treatment, and as surrogate outcomes, i.e. in place of clinical outcomes in trials [118]. Current development *methods* tend to determine drug development programs rather than the characteristics of the particular drugs in question. There is both substantial hope and still underappreciated challenges ahead.

References

1 Schneider LS: Issues in design and conduct of clinical trials for cognitive-enhancing drugs; in McArthur RA, Borsini F (eds): Animal and Translational Models for CNS Drug Discovery. San Diego, Academic Press, 2008, pp 21–76.

2 Lindner MD, McArthur RA, Deadwyler S, Hampson R, Tariot PN: Development, optimization and use of preclinical behavioral models to maximize the productivity of drug discovery for Alzheimer's disease; in McArthur RA, Borsini F (eds): Animal and Translational Models for CNS Drug Discovery. San Diego, Academic Press, 2008, pp 93–157.

3 Bartus R, Dean R, Beer B, Lippa A: The cholinergic hypothesis of geriatric memory dysfunction. Science 1982;217:408–414.

4 Schneider LS: Antidementia drugs; in Sadock BJ, Sadock MD, Ruiz P (eds): Comprehensive Textbook of Psychiatry, ed 9. Philadelphia, Lippincott Williams & Wilkins, 2009, vol 2, pp 4119–4130.

5 McArthur RA, Gray J, Schreiber R: Cognitive effects of muscarinic M_1 functional agonists in non-human primates and clinical trials. Curr Opin Invest Drugs 2010;11:740–760.

6 Birks J, Harvey RJ: Donepezil for dementia due to Alzheimer's disease. Cochrane Database Syst Rev 2006:CD001190.

7 Courtney C, Farrell D, Gray R, Hills R, Lynch L, Sellwood E, Edwards S, Hardyman W, Raftery J, Crome P, Lendon C, Shaw H, Bentham P: Long-term donepezil treatment in 565 patients with Alzheimer's disease (AD2000): randomised double-blind trial. Lancet 2004;363:2105–2115.

8 Schneider LS, Anand R, Farlow M: Systematic review of the efficacy of rivastigmine for the patients with Alzheimer's disease. Int J Geriatr Psychopharmacol 1998;1:S26–S34.

9 Birks J, Grimley Evans J, Iakovidou V, Tsolaki M: Rivastigmine for Alzheimer's disease. Cochrane Database Syst Rev 2009:CD001191.

10 Corey-Bloom J, Anand R, Veach J: A randomized trial evaluating the efficacy and safety of ENA 713 (rivastigmine tartrate), a new acetylcholinesterase inhibitor, in patients with mild to moderately severe Alzheimer's disease. Int J Geriatr Psychopharmacol 1998;1:55–65.

11 Rosler M, Anand R, Cicin-Sain A, Gauthier S, Agid Y, Dal-Bianco P, Stahelin HB, Hartman R, Gharabawi M: Efficacy and safety of rivastigmine in patients with Alzheimer's disease: international randomised controlled trial. BMJ 1999;318:633–638.

12 Winblad B, Grossberg G, Frölich L, Farlow M, Zechner S, Nagel J, Lane R: Ideal. Neurology 2007;69:S14–S22.

13 Loy C, Schneider L: Galantamine for Alzheimer's disease and mild cognitive impairment. Cochrane Database Syst Rev 2006:CD001747.

14 Schneider L, Tariot P: Cognitive enhancers and treatments for Alzheimer's disease; in Tasman A, Kay J, Lieberman J (eds): Psychiatry. London, Wiley, 2004, pp 2096–2108.
15 Gill SS, Anderson GM, Fischer HD, Bell CM, Li P, Normand S-LT, Rochon PA: Syncope and its consequences in patients with dementia receiving cholinesterase inhibitors: a population-based cohort study. Arch Intern Med 2009;169:867–873.
16 Birks J: Cholinesterase inhibitors for Alzheimer's disease. Cochrane Database Syst Rev 2006:CD005593.
17 Schmitt FA, Ashford W, Ernesto C, Saxton J, Schneider LS, Clark CM, Ferris SH, Mackell JA, Schafer K, Thal LJ: The Severe Impairment Battery: concurrent validity and the assessment of longitudinal change in Alzheimer's disease. Alzheimer Dis Assoc Disord 1997;11:S51–S56.
18 Farlow MR, Salloway S, Tariot PN, Yardley J, Moline ML, Wang Q, Brand-Schieber E, Zou H, Hsu T, Satlin A: Effectiveness and tolerability of high-dose (23 mg/day) versus standard-dose (10 mg/day) donepezil in moderate to severe Alzheimer's disease: a 24-week, randomized, double-blind study. Clin Ther 2010;32:1234–1251.
19 Winblad B, Gauthier S, Scinto L, Feldman H, Wilcock GK, Truyen L, Mayorga AJ, Wang D, Brashear HR, Nye JS: Safety and efficacy of galantamine in subjects with mild cognitive impairment. Neurology 2008;70:2024–2035.
20 Feldman HH, Ferris S, Winblad B, Sfikas N, Mancione L, He Y, Tekin S, Burns A, Cummings J, del Ser T, Inzitari D, Orgogozo J-M, Sauer H, Scheltens P, Scarpini E, Herrmann N, Farlow M, Potkin S, Charles HC, Fox NC, Lane R: Effect of rivastigmine on delay to diagnosis of Alzheimer's disease from mild cognitive impairment: the InDDEx study. Lancet Neurol 2007;6:501–512.
21 Petersen RC, Thomas RG, Grundman M, Bennett D, Doody R, Ferris S, Galasko D, Jin S, Kaye J, Levey A, Pfeiffer E, Sano M, van Dyck CH, Thal LJ, Alzheimer's Disease Cooperative Study G: Vitamin E and donepezil for the treatment of mild cognitive impairment. N Engl J Med 2005;352:2379–2388.
22 Salloway S, Ferris S, Kluger A, Goldman R, Griesing T, Kumar D, Richardson S, Donepezil 401 Study G: Efficacy of donepezil in mild cognitive impairment: a randomized placebo-controlled trial. Neurology 2004;63:651–657.
23 Raschetti R, Albanese E, Vanacore N, Maggini M: Cholinesterase inhibitors in mild cognitive impairment: a systematic review of randomised trials. PLoS Med 2007;4:e338.
24 Birks J, Flicker L: Donepezil for mild cognitive impairment. Cochrane Database Syst Rev 2006: CD006104.
25 Reisberg B, Doody R, Stoffler A, Schmitt F, Ferris S, Mobius HJ: Memantine in moderate-to-severe Alzheimer's disease. N Engl J Med 2003;348: 1333–1341.
26 Tariot PN, Farlow MR, Grossberg GT, Graham SM, McDonald S, Gergel I: Memantine treatment in patients with moderate to severe Alzheimer disease already receiving donepezil: a randomized controlled trial. JAMA 2004;291:317–324.
27 Van Dyck CH, Tariot PN, Meyers B, Resnick ME: A 24-week randomized, controlled trial of memantine in patients with moderate-to-severe Alzheimer disease. Alzheimer Dis Assoc Disord 2007;21: 136–143.
28 Winblad B, Poritis N: Memantine in severe dementia: results of the 9M-BEST study (benefit and efficacy in severely demented patients during treatment with memantine). Int J Geriatr Psychiatry 1999;14: 135–146.
29 Peskind ER, Potkin SG, Pomara N, Ott BR, Graham SM, Olin JT, McDonald S: Memantine treatment in mild to moderate Alzheimer disease: a 24-week randomized, controlled trial. Am J Geriatr Psychiatry 2006;14:704–715.
30 Bakchine S, Loft H: Memantine treatment in patients with mild to moderate Alzheimer's disease: results of a randomised, double-blind, placebo-controlled 6-month study. J Alzheimers Dis 2008;13: 97–107.
31 Porsteinsson AP, Grossberg GT, Mintzer J, Olin JT: Memantine treatment in patients with mild to moderate Alzheimer's disease already receiving a cholinesterase inhibitor: a randomized, double-blind, placebo-controlled trial. Curr Alzheimer Res 2008; 5:83–89.
32 Winblad B, Jones RW, Wirth Y, Stöffler A, Möbius HJ: Memantine in moderate to severe Alzheimer's disease: A meta-analysis of randomised clinical trials. Dementia and Geriatric Cognitive Disorders 2007;24:20–27.
33 Schneider LS, Dagerman KS, Higgins JPT, McShane R: Lack of evidence for the efficacy of memantine in mild Alzheimer disease. Arch Neurol 2011;68: 991–998.
34 McShane R, Areosa Sastre A, Minakaran N: Memantine for dementia. Cochrane Database Syst Rev 2006:CD003154.
35 Frankiewicz T, Parsons CG: Memantine restores long-term potentiation impaired by tonic N-methyl-D-aspartate receptor activation following reduction of Mg^{2+} in hippocampal slices. Neuropharmacology 1999;38:1253–1259.

36 Martinez-Coria H, Green KN, Billings LM, Kitazawa M, Albrecht M, Rammes G, Parsons CG, Gupta S, Banerjee P, LaFerla FM: Memantine improves cognition and reduces Alzheimer's-like neuropathology in transgenic mice. Am J Pathol 2010;176:870–880.

37 Lopez OL, Becker JT, Wahed AS, Saxton J, Sweet RA, Wolk DA, Klunk W, Dekosky ST: Long-term effects of the concomitant use of memantine with cholinesterase inhibition in Alzheimer disease. J Neurol Neurosurg Psychiatry 2009;80:600–607.

38 Rountree SD, Chan W, Pavlik VN, Darby EJ, Siddiqui S, Doody RS: Persistent treatment with cholinesterase inhibitors and/or memantine slows clinical progression of Alzheimer disease. Alzheimers Res Ther 2009;1:7.

39 Reisberg B, Doody R, Stoffler A, Schmitt F, Ferris S, Mobius HJ: A 24-week open-label extension study of memantine in moderate to severe Alzheimer disease. Arch Neurol 2006;63:49–54.

40 Schneider LS: Open-label extension studies and misinformation. Arch Neurol 2006;63:1036.

41 Schneider LS, Sano M: Current Alzheimer's disease clinical trials: methods and placebo outcomes. Alzheimers Dement 2009;5:388–397.

42 Alvarez X, Cacabelos R, Sampedro C, Couceiro V, Aleixandre M, Vargas M, Linares C, Granizo E, Garcia-Fantini M, Baurecht W, Doppler E, Moessler H: Combination treatment in Alzheimer's disease: results of a randomized, controlled trial with cerebrolysin and donepezil. Curr Alzheimer Res 2011; 8:583–591.

43 Schneider LS: *Ginkgo biloba* extract and preventing Alzheimer disease. JAMA 2008;300:2306–2308.

44 Birks J, Grimley Evans J: *Ginkgo biloba* for Cognitive Impairment and Dementia. Chichester, Wiley, 2009.

45 Schneider LS, DeKosky ST, Farlow MR, Tariot PN, Hoerr R, Kieser M: A randomized, double-blind, placebo-controlled trial of two doses of *Ginkgo biloba* extract in dementia of the Alzheimer's type. Curr Alzheimer Res 2005;2:541–551.

46 McCarney R, Fisher P, Iliffe S, van Haselen R, Griffin M, van der Meulen J, Warner J: *Ginkgo biloba* for mild to moderate dementia in a community setting: a pragmatic, randomised, parallel-group, double-blind, placebo-controlled trial. Int J Geriatr Psychiatry 2008;23:1222–1230.

47 DeKosky ST, Williamson JD, Fitzpatrick AL, Kronmal RA, Ives DG, Saxton JA, Lopez OL, Burke G, Carlson MC, Fried LP, Kuller LH, Robbins JA, Tracy RP, Woolard NF, Dunn L, Snitz BE, Nahin RL, Furberg CD, for the Ginkgo Evaluation of Memory Study Investigators: *Ginkgo biloba* for prevention of dementia: a randomized controlled trial. JAMA 2008;300:2253–2262.

48 Vellas B, Andrieu S, Ousset PJ, Ouzid M, Mathiex-Fortunet H, for the GuidAge Study G: The GuidAge study: Methodological issues. A 5-year double-blind randomized trial of the efficacy of EGb 761 for prevention of Alzheimer disease in patients over 70 with a memory complaint. Neurology 2006;67: S6–S11.

49 Dodge HH, Zitzelberger T, Oken BS, Howieson D, Kaye J: A randomized placebo-controlled trial of *Ginkgo biloba* for the prevention of cognitive decline. Neurology 2008;70:1809–1817.

50 Kaduszkiewicz H, Zimmermann T, Beck-Bornholdt H-P, van den Bussche H: Cholinesterase inhibitors for patients with Alzheimer's disease: systematic review of randomised clinical trials. BMJ 2005;331: 321–327.

51 Raina P, Santaguida P, Ismaila A, Patterson C, Cowan D, Levine M, Booker L, Oremus M: Effectiveness of cholinesterase inhibitors and memantine for treating dementia: evidence review for a clinical practice guideline. Ann Intern Med 2008;148:379–397.

52 Takeda A, Loveman E, Clegg A, Kirby J, Picot J, Payne E, Green C: A systematic review of the clinical effectiveness of donepezil, rivastigmine and galantamine on cognition, quality of life and adverse events in Alzheimer's disease. Int J Geriatr Psychiatry 2006;21:17–28.

53 Loveman E, Green C, Kirby J, Takeda A, Picot J, Payne E, Clegg A: The clinical and cost-effectiveness of donepezil, rivastigmine, galantamine and memantine for Alzheimer's disease. Health Technol Assess 2006;10:iii–iv, ix–xi, 1–160.

54 Rockwood K, Fay S, Song X, MacKnight C, Gorman M: Video-imaging synthesis of treating Alzheimer's disease. I. Attainment of treatment goals by people with Alzheimer's disease receiving galantamine: a randomized controlled trial. CMAJ 2006;174: 1099–1105.

55 Winblad B, Engedal K, Soininen H, Verhey F, Waldemar G, Wimo A, Wetterholm AL, Zhang R, Haglund A, Subbiah P, Donepezil Nordic Study G: A 1-year, randomized, placebo-controlled study of donepezil in patients with mild to moderate AD. Neurology 2001;57:489–495.

56 Lopez OL, Becker JT, Wahed AS, Saxton J, Sweet RA, Wolk DA, Klunk W, DeKosky ST: Long-term effects of the concomitant use of memantine with cholinesterase inhibition in Alzheimer disease. J Neurol Neurosurg Psychiatry 2009;80:600–607.

57 Atri A, Shaughnessy L, Locascio JJ, Growdon JH: Long-term course and effectiveness of combination therapy in Alzheimer disease. Alzheimer Dis Assoc Disord 2008;22:209–221.

58 Geldmacher DS, Provenzano G, McRae T, Mastey V, Ieni JR: Donepezil is associated with delayed nursing home placement in patients with Alzheimer's disease. J Am Geriatr Soc 2003;51:937–944.
59 Rountree S, Chan W, Pavlik V, Darby E, Siddiqui S, Doody R: Persistent treatment with cholinesterase inhibitors and/or memantine slows clinical progression of Alzheimer disease. Alzheimers Res Ther 2009;1:7.
60 Schneider LS, Qizilbash N: Delay in nursing home placement with donepezil. J Am Geriatr Soc 2004; 52:1024–1026.
61 Schneider LS, Insel PS, Weiner MW, for the Alzheimer's Disease Neuroimaging Initiative: Treatment with cholinesterase inhibitors and memantine of patients in the Alzheimer's disease neuroimaging initiative. Arch Neurol 2011;68: 58–66.
62 Sona A, Zhang P, Ames D, Bush AI, Lautenschlager NT, Martins RN, Masters CL, Rowe CC, Szoeke C, Taddei K, Ellis KA: Predictors of rapid cognitive decline in Alzheimer's disease: results from the Australian Imaging, Biomarkers and Lifestyle (AIBL) study of ageing. Int Psychogeriatr 2011;24: 197–204.
63 Holmes C, Wilkinson D, Dean C, Vethanayagam S, Olivieri S, Langley A, Pandita-Gunawardena ND, Hogg F, Clare C, Damms J: The efficacy of donepezil in the treatment of neuropsychiatric symptoms in Alzheimer disease. Neurology 2004;63:214–219.
64 Mauskopf JA, Paramore C, Lee WC, Snyder EH: Drug persistency patterns for patients treated with rivastigmine or donepezil in usual care settings. J Manag Care Pharm 2005;11:231–251.
65 Rodda J, Morgan S, Walker Z: Are cholinesterase inhibitors effective in the management of the behavioral and psychological symptoms of dementia in Alzheimer's disease? A systematic review of randomized, placebo-controlled trials of donepezil, rivastigmine and galantamine. Int Psychogeriatr 2009;21:813–824.
66 Campbell N, Ayub A, Boustani MA, Fox C, Farlow M: Impact of cholinesterase inhibitors on behavioral and psychological symptoms of Alzheimer's disease: a meta-analysis. Clin Interv Aging 2008;3: 719–728.
67 Howard RJ, Juszczak E, Ballard CG, Bentham P, Brown RG, Bullock R, Burns AS, Holmes C, Jacoby R, Johnson T, Knapp M, Lindesay J, O'Brien JT, Wilcock G, Katona C, Jones RW, DeCesare J, Rodger M, the C-ADTG: Donepezil for the treatment of agitation in Alzheimer's disease. N Engl J Med 2007; 357:1382–1392.
68 Schneider LS, Dagerman KS, Insel P: Risk of death with atypical antipsychotic drug treatment for dementia: Meta-analysis of randomized placebo-controlled trials. JAMA 2005;294:1934–1943.
69 Ballard C, Hanney ML, Theodoulou M, Douglas S, McShane R, Kossakowski K, Gill R, Juszczak E, Yu LM, Jacoby R: The Dementia Antipsychotic Withdrawal Trial (DART-AD): long-term follow-up of a randomised placebo-controlled trial. Lancet Neurol 2009;8:151–157.
70 Kales HC, Valenstein M, Kim HM, McCarthy JF, Ganoczy D, Cunningham F, Blow FC: Mortality risk in patients with dementia treated with antipsychotics versus other psychiatric medications. Am J Psychiatry 2007;164:1568–1576.
71 Kales HC, Zivin K, Kim HM, Valenstein M, Chiang C, Ignacio R, Ganoczy D, Cunningham F, Schneider LS, Blow FC: Trends in antipsychotic use in dementia 1999–2007. Arch Gen Psychiatry 2011;68: 190–197.
72 Rosenberg PB, Drye LT, Martin BK, Frangakis C, Mintzer JE, Weintraub D, Porsteinsson AP, Schneider LS, Rabins PV, Munro CA, Meinert CL, Lyketsos CG, Group FtD-R: Sertraline for the treatment of depression in Alzheimer disease. Am J Geriatr Psychiatry 2010;18:136–145.
73 Weintraub D, Rosenberg PB, Drye LT, Martin BK, Frangakis C, Mintzer JE, Porsteinsson AP, Schneider LS, Rabins PV, Munro CA, Meinert CL, Lyketsos CG, Group FtD-R: Sertraline for the treatment of depression in Alzheimer disease: week-24 outcomes. Am J Geriatr Psychiatry 2010;18:332–340.
74 Olin JT, Katz IR, Meyers BS, Schneider LS, Lebowitz BD: Provisional diagnostic criteria for depression of Alzheimer disease: rationale and background. Am J Geriatr Psychiatry 2002;10:129–141.
75 Banerjee S, Hellier J, Dewey M, Romeo R, Ballard C, Baldwin R, Bentham P, Fox C, Holmes C, Katona C, Knapp M, Lawton C, Lindesay J, Livingston G, McCrae N, Moniz-Cook E, Murray J, Nurock S, Orrell M, O'Brien J, Poppe M, Thomas A, Walwyn R, Wilson K, Burns A: Sertraline or mirtazapine for depression in dementia (HTA-SADD): a randomised, multicentre, double-blind, placebo-controlled trial. Lancet 2011;378:403–411.
76 Pollock BG, Mulsant BH, Rosen J, Sweet RA, Mazumdar S, Bharucha A, Marin R, Jacob NJ, Huber KA, Kastango KB, Chew ML: Comparison of citalopram, perphenazine, and placebo for the acute treatment of psychosis and behavioral disturbances in hospitalized, demented patients. Am J Psychiatry 2002;159:460–465.
77 Martinón-Torres G, Fioravanti M, Grimley Evans J: Trazodone for agitation in dementia. Cochrane Database Syst Rev 2004:CD004990.

78 Teri L, Logsdon RG, Peskind E, Raskind M, Weiner MF, Tractenberg RE, Foster NL, Schneider LS, Sano M, Whitehouse P, Tariot P, Mellow AM, Auchus AP, Grundman M, Thomas RG, Schafer K, Thal LJ, Alzheimer's Disease Cooperative S: Treatment of agitation in AD: a randomized, placebo-controlled clinical trial. Neurology 2000;55:1271–1278.

79 Becker RE, Greig NH, Giacobini E: Why do so many drugs for Alzheimer's disease fail in development? Time for new methods and new practices? J Alzheimers Dis 2008;15:303–325.

80 Paul SM, Mytelka DS, Dunwiddie CT, Persinger CC, Munos BH, Lindborg SR, Schacht AL: How to improve R&D productivity: the pharmaceutical industry's grand challenge. Nat Rev Drug Discov 2010;9:203–214.

81 Mohs RC, Knopman D, Petersen RC, Ferris SH, Ernesto C, Grundman M, Sano M, Bieliauskas L, Geldmacher D, Clark C, Thal LJ: Development of cognitive instruments for use in clinical trials of antidementia drugs: additions to the Alzheimer's disease assessment scale that broaden its scope. The Alzheimer's Disease Cooperative Study. Alzheimer Dis Assoc Disord 1997;11:S13–S21.

82 Galasko D, Bennett D, Sano M, Ernesto C, Thomas R, Grundman M, Ferris S: An inventory to assess activities of daily living for clinical trials in Alzheimer's disease. Alzheimer Dis Assoc Disord 1997;11:33–39.

83 Schneider LS, Olin JT, Doody RS, Clark CM, Morris JC, Reisberg B, Schmitt FA, Grundman M, Thomas RG, Ferris SH: Validity and reliability of the Alzheimer's Disease Cooperative Study – Clinical Global Impression of Change. The Alzheimer's Disease Cooperative Study. Alzheimer Dis Assoc Disord 1997;11:S22–S32.

84 Morris JC: The Clinical Dementia Rating: current version and scoring rules. Neurology 1993;43: 2412–2414.

85 Holmes C, Boche D, Wilkinson D, Yadegarfar G, Hopkins V, Bayer A, Jones RW, Bullock R, Love S, Neal JW, Zotova E, Nicoll JAR: Long-term effects of $A\beta_{42}$ immunisation in Alzheimer's disease: follow-up of a randomised, placebo-controlled phase I trial. Lancet 2008;372:216–223.

86 Rinne JO, Brooks DJ, Rossor MN, Fox NC, Bullock R, Klunk WE, Mathis CA, Blennow K, Barakos J, Okello AA, de Liano SRM, Liu E, Koller M, Gregg KM, Schenk D, Black R, Grundman M: ^{11}C-PIB PET assessment of change in fibrillar amyloid-β load in patients with Alzheimer's disease treated with bapineuzumab: a phase 2, double-blind, placebo-controlled, ascending-dose study. Lancet Neurol 2010;9:363–372.

87 Gauthier S, Aisen P, Ferris S, Saumier D, Duong A, Haine D, Garceau D, Suhy J, Oh J, Lau W, Sampalis J: Effect of tramiprosate in patients with mild-to-moderate Alzheimer's disease: exploratory analyses of the MRI subgroup of the Alphase Study. J Nutr Health Aging 2009;13:550–557.

88 Ostrowitzki S, Deptula D, Thurfjell L, Barkhof F, Bohrmann B, Brooks DJ, Klunk WE, Ashford E, Yoo K, Xu Z-X, Loetscher H, Santarelli L: Mechanism of amyloid removal in patients with Alzheimer disease treated with gantenerumab. Arch Neurol 2011 (E-pub ahead of print).

89 Golde TE, Schneider LS, Koo EH: Anti-aβ therapeutics in Alzheimer's disease: the need for a paradigm shift. Neuron 2011;69:203–213.

90 Hardy J, Selkoe DJ: The amyloid hypothesis of Alzheimer's disease: progress and problems on the road to therapeutics. Science 2002;297:353–356.

91 Chen H-K, Ji Z-S, Dodson SE, Miranda RD, Rosenblum CI, Reynolds IJ, Freedman SB, Weisgraber KH, Huang Y, Mahley RW: Apolipoprotein ε4 domain interaction mediates detrimental effects on mitochondria and is a potential therapeutic target for Alzheimer's disease. J Biol Chem 2011;286: 5215–5221.

92 Gilman S, Koller M, Black RS, Jenkins L, Griffith SG, Fox NC, Eisner L, Kirby L, Rovira MB, Forette F, Orgogozo JM: Clinical effects of Aβ immunization (AN1792) in patients with AD in an interrupted trial. Neurology 2005;64:1553–1562.

93 Castellano JM, Kim J, Stewart FR, Jiang H, DeMattos RB, Patterson BW, Fagan AM, Morris JC, Mawuenyega KG, Cruchaga C, Goate AM, Bales KR, Paul SM, Bateman RJ, Holtzman DM: Human ApoE isoforms differentially regulate brain amyloid-β peptide clearance. Science Translat Med 2011;3:89ra57.

94 Harrison J, Minassian SL, Jenkins L, Black RS, Koller M, Grundman M: A Neuropsychological Test Battery for use in Alzheimer disease clinical trials. Arch Neurol 2007;64:1323–1329.

95 Orgogozo J-M, Gilman S, Dartigues J-F, Laurent B, Puel M, Kirby LC, Jouanny P, Dubois B, Eisner L, Flitman S, Michel BF, Boada M, Frank A, Hock C: Subacute meningoencephalitis in a subset of patients with AD after $A\beta_{42}$ immunization. Neurology 2003; 61:46–54.

96 Muhs A, Hickman DT, Pihlgren M, Chuard N, Giriens V, Meerschman C, van der Auwera I, van Leuven F, Sugawara M, Weingertner M-C, Bechinger B, Greferath R, Kolonko N, Nagel-Steger L, Riesner D, Brady RO, Pfeifer A, Nicolau C: Liposomal vaccines with conformation-specific amyloid peptide antigens define immune response and efficacy in APP transgenic mice. Proc Natl Acad Sci USA 2007; 104:9810–9815.

97 Siemers ER, Friedrich S, Dean RA, Gonzales CR, Farlow MR, Paul SM, DeMattos RB: Safety and changes in plasma and cerebrospinal fluid amyloid-β after a single administration of an amyloid-β monoclonal antibody in subjects with Alzheimer disease. Clin Neuropharmacol 2010;33:67–73.
98 Salloway S, Sperling R, Gilman S, Fox NC, Blennow K, Raskind M, Sabbagh M, Honig LS, Doody R, van Dyck CH, Mulnard R, Barakos J, Gregg KM, Liu E, Lieberburg I, Schenk D, Black R, Grundman M, Bapineuzumab 201 Clinical Trial Investigators: A phase 2 multiple ascending dose trial of bapineuzumab in mild to moderate Alzheimer disease. Neurology 2009;73:2061–2070.
99 Dodel R, Neff F, Noelker C, Pul R, Du Y, Bacher M, Oertel W: Intravenous immunoglobulins as a treatment for Alzheimer's disease: rationale and current evidence. Drugs 2010;70:513–528.
100 Wilcock GK, Black SE, Hendrix SB, Zavitz KH, Swabb EA, Laughlin MA: Efficacy and safety of tarenflurbil in mild to moderate Alzheimer's disease: a randomised phase II trial. Lancet Neurol 2008;7:483–493.
101 Green RC, Schneider LS, Hendrix S, Zavitz K, Swabb E: Safety and efficacy of tarenflurbil in subjects with mild Alzheimer's disease: results from an 18-month multicenter phase 3 trial late breaking news. Eur J Neurol 2008;15:412–412.
102 May PC, Dean RA, Lowe SL, Martenyi F, Sheehan SM, Boggs LN, Monk SA, Mathes BM, Mergott DJ, Watson BM, Stout SL, Timm DE, Smith LaBell E, Gonzales CR, Nakano M, Jhee SS, Yen M, Ereshefsky L, Lindstrom TD, Calligaro DO, Cocke PJ, Greg Hall D, Friedrich S, Citron M, Audia JE: Robust central reduction of amyloid-β in humans with an orally available, non-peptidic β-secretase inhibitor. J Neurosci 2011;31:16507–16516.
103 Aisen PS, Gauthier S, Vellas B, Briand R, Saumier D, Laurin J, Garceau D: Alzhemed: a potential treatment for Alzheimer's disease. Curr Alzheimer Res 2007;4:473–478.
104 Aisen PS, Saumier D, Briand R, Laurin J, Gervais F, Tremblay P, Garceau D: A phase II study targeting amyloid-β with 3APS in mild-to-moderate Alzheimer disease. Neurology 2006;67:1757–1763.
105 McLaurin J, Kierstead ME, Brown ME, Hawkes CA, Lambermon MHL, Phinney AL, Darabie AA, Cousins JE, French JE, Lan MF, Chen F, Wong SSN, Mount HTJ, Fraser PE, Westaway D, George-Hyslop PS: Cyclohexanehexol inhibitors of Aβ aggregation prevent and reverse Alzheimer phenotype in a mouse model. Nat Med 2006;12:801–808.
106 Salloway S, Sperling R, Keren R, Porsteinsson AP, van Dyck CH, Tariot PN, Gilman S, Arnold D, Abushakra S, Hernandez C, Crans G, Liang E, Quinn G, Bairu M, Pastrak A, Cedarbaum JM, ELND005-AD201 Investigators: A phase 2 randomized trial of ELND005, scyllo-inositol, in mild to moderate Alzheimer disease. Neurology 2011;77:1253–1262.
107 Lannfelt L, Blennow K, Zetterberg H, Batsman S, Ames D, Harrison J, Masters CL, Targum S, Bush AI, Murdoch R, Wilson J, Ritchie CW: Safety, efficacy, and biomarker findings of PBT-2 in targeting Aβ as a modifying therapy for Alzheimer's disease: a phase IIA, double-blind, randomised, placebo-controlled trial. Lancet Neurol 2008;7:779–786.
108 Blessed G, Tomlinson BE, Roth M: The association between quantitative measures of dementia and of senile change in the cerebral grey matter of elderly subjects. Br J Psychiatry 1968;114:797–811.
109 Braak H, Del Tredici K: Alzheimer's pathogenesis: Is there neuron-to-neuron propagation? Acta Neuropathol (Berl) 2011;121:589–595.
110 Boutajangout A, Sigurdsson EM, Krishnamurthy PK: Tau as a therapeutic target for Alzheimer's disease. Curr Alzheimer Res 2011;8:666–677.
111 Fuentes P, Catalan J: A clinical perspective: anti-tau's treatment in Alzheimer's disease. Curr Alzheimer Res 2011;8:686–688.
112 Hampel H, Ewers M, Berger K, Annas P: Lithium trial in Alzheimer's disease: a randomized, single-blind, placebo-controlled, multicenter 10-week study. J Clin Psychiatry 2009;70:922–931.
113 Forlenza OV, Diniz BS, Radanovic M, Santos FS, Talib LL, Gattaz WF: Disease-modifying properties of long-term lithium treatment for amnestic mild cognitive impairment: randomised controlled trial. Br J Psychiatry 2011;198:351–356.
114 Tariot PN, Schneider LS, Cummings J, Thomas RG, Raman R, Jakimovich LJ, Loy R, Bartocci B, Fleisher A, Ismail MS, Porsteinsson A, Weiner M, Jack CR Jr, Thal L, Aisen PS, for the Alzheimer's Disease Cooperative Study Group: Chronic divalproex sodium to attenuate agitation and clinical progression of Alzheimer disease. Arch Gen Psychiatry 2011;68:853–861.
115 Lasagna-Reeves CA, Castillo-Carranza DL, Jackson GR, Kayed R: Tau oligomers as potential targets for immunotherapy for Alzheimer's disease and tauopathies. Curr Alzheimer Res 2011;8:659–665.
116 Maher-Edwards G, Zvartau-Hind M, Hunter AJ, Gold M, Hopton G, Jacobs G, Davy M, Williams P: Double-blind, controlled phase II study of a 5-HT_6 receptor antagonist, SB-742457, in Alzheimer's disease. Curr Alzheimer Res 2010;7:374–385.

117 Maher-Edwards G, Dixon R, Hunter J, Gold M, Hopton G, Jacobs G, Hunter J, Williams P: SB-742457 and donepezil in Alzheimer disease: A randomized, placebo-controlled study. Int J Geriatr Psychiatry 2011;26:536–544.

118 Hampel H, Frank R, Broich K, Teipel SJ, Katz RG, Hardy J, Herholz K, Bokde ALW, Jessen F, Hoessler YC, Sanhai WR, Zetterberg H, Woodcock J, Blennow K: Biomarkers for Alzheimer's disease: academic, industry and regulatory perspectives. Nat Rev Drug Discov 2010;9:560–574.

119 Meinert CL, McCaffrey LD, Breitner JC: Alzheimer's disease anti-inflammatory prevention trial: design, methods, and baseline results. Alzheimers Dement 2009;5:93–104.

120 Snitz BE, O'Meara ES, Carlson MC, Arnold AM, Ives DG, Rapp SR, Saxton J, Lopez OL, Dunn LO, Sink KM, DeKosky ST: *Ginkgo biloba* for preventing cognitive decline in older adults: a randomized trial. JAMA 2009;302:2663–2670.

121 Vellas B, Andrieu S, Ousset PJ, Ouzid M, Mathiex-Fortunet H: The GuidAge study: methodological issues. A 5-year double-blind randomized trial of the efficacy of EGb 761 for prevention of Alzheimer disease in patients over 70 with a memory complaint. Neurology 2006;67:S6–S11.

122 Christen WG, Gaziano JM, Hennekens CH: Design of physicians' health study II – a randomized trial of β-carotene, vitamins E and C, and multivitamins, in prevention of cancer, cardiovascular disease, and eye disease, and review of results of completed trials. Ann Epidemiol 2000;10:125–134.

123 Heart Protection Study Collaborative Group: MRC/BHF Heart Protection Study of cholesterol lowering with simvastatin in 20,536 high-risk individuals: a randomised placebo-controlled trial. Lancet 2002; 360:7–22.

124 Heart Protection Study Collaborative Group: MRC/BHF Heart Protection Study of antioxidant vitamin supplementation in 20,536 high-risk individuals: a randomised placebo-controlled trial. Lancet 2002; 360:23–33.

125 Kryscio RJ, Mendiondo MS, Schmitt FA, Markesbery WR: Designing a large prevention trial: statistical issues. Stat Med 2004;23:285–296.

126 Grady D, Yaffe K, Kristof M, Lin F, Richards C, Barrett-Connor E: Effect of postmenopausal hormone therapy on cognitive function: the Heart and Estrogen/Progestin Replacement Study. Am J Med 2002;113:543–548.

127 Craig MC, Maki PM, Murphy DG: The Women's Health Initiative Memory Study: findings and implications for treatment. Lancet Neurol 2005;4:190–194.

Prof. Lon S. Schneider, MD
University of Southern California Keck School of Medicine
Los Angeles, CA 90033 (USA)
Tel. +1 323 442 7600
E-Mail lschneid@usc.edu

Hampel H, Carrillo MC (eds): Alzheimer's Disease – Modernizing Concept, Biological Diagnosis and Therapy.
Adv Biol Psychiatry. Basel, Karger, 2012, vol 28, pp 168–178

Regulatory Requirements on Clinical Trials in Alzheimer's Disease

Karl Broich[a] · Gabriele Schlosser-Weber[a] · Marco Weiergräber[a] · Harald Hampel[b]

[a]Federal Institute for Drugs and Medical Devices (BfArM), Bonn, and [b]Department of Psychiatry, Psychosomatic Medicine and Psychotherapy, Goethe University, Frankfurt a.M., Germany

Abstract

Recent progress in understanding the neurobiology and pathophysiology of Alzheimer's disease (AD) has fostered new interest for more efficacious symptomatic treatments as well as for disease-modifying approaches based on these insights. This has many implications for regulatory scientific advice and decision-making. Leaving the concept of the full picture of dementia as a first step of diagnosis for AD requires new standardization of diagnostic criteria, particularly for earlier disease stages. In consequence, new validated assessment tools sensitive to change for the different dimensions of AD are necessary in early AD. If a treatment claim for disease modification is strived for, it has to be shown that the treatment has an impact on the underlying pathophysiology of AD in addition to clinical improvement. This will lead to more complex study designs (adaptive designs, staggered start and withdrawal designs, combination of symptomatic and potential disease-modifying compounds, etc.). In addition to this, progress in qualification and validation of biomarkers might play an important role in future drug development for AD. Some biomarkers are already in the process of implementation as outcome variables into regulatory guideline documents, e.g. regarding phase II in drug development programs as outcome measures in proof-of-concept or dose-finding studies. Further validation of specific biomarkers in large-scale international controlled multicenter studies is necessary before they can be accepted as outcome measures in pivotal phase III clinical trials. Unfortunately, no biomarker has been sufficiently validated to be acceptable as a surrogate endpoint. Establishing surrogate endpoints is an important goal in early AD as traditional clinical outcome measures might be too subtle to be sensitive to change or need unfeasible treatment durations for clinical trial conditions. Improvements can only be accomplished by active synergistic collaboration between academic, industrial and regulatory partners.

Despite rapid progress in understanding the neurobiology and pathophysiology of Alzheimer's disease (AD), only several cholinesterase inhibitors and memantine have been approved for symptomatic treatment with overall limited symptomatic

The opinions expressed are those of the authors and do not necessarily reflect the views of the BfArM.

improvement, whereas for example in Parkinson's disease, longer lasting symptomatic improvement effects are much more impressive compared to AD, yet no product has been approved for disease modification in any neurodegenerative disorder by regulatory bodies. Unfortunately, many recent clinical trials have failed to show symptomatic improvement or promising hints for disease modification in AD, e.g. medicinal products with an Aβ-targeted mode of action. Several reasons must be considered for these negative outcomes: the amyloid cascade hypothesis might be more important for the very early disease stages and not for patients with fully established features of dementia. Patient populations included in the clinical trials still showed a certain amount of heterogeneity and medicinal products might have been introduced to phase III programs without proper phase II testing [1–5].

The term dementia describes a syndrome characterized by memory impairment, intellectual deterioration, changes in personality and behavioral abnormalities (DSM-IV-TR, ICD-10). These symptoms are of significant severity to interfere with social activities and occupational functioning. Moreover, the observed cognitive deficits must represent a decline from a higher level of function. In general, the disorders constituting the dementia syndromes share a common symptom presentation and are identified and classified on the basis of different etiologies and pathophysiologies. However, there is now a common understanding between all stakeholders that the full picture of dementia represents an advanced disease stage, which limits the treatment options with a chance of clinically meaningful improvements. This underscores the need for new criteria of AD and its earlier disease stages.

The main goals of treatments for AD might be: (1) symptomatic improvement, which may consist mainly in enhanced cognition and improved daily functioning, more autonomy and/or improvement in neuropsychiatric and behavioral dysfunction; (2) disease modification with slowing or arrest of symptom progression of the dementing process, and (3) primary prevention of disease by intervention in key pathogenic mechanisms at a presymptomatic stage.

For symptomatic treatment the development and use of relevant reliable and sensitive instruments to measure cognition, functional and behavioral symptoms, particularly for the assessment of activities of daily living and behavioral symptoms is encouraged.

Currently there is a lack of agreement on the appropriate methodology to demonstrate slowing or arrest of the dementing process [6, 7]. Data on prevention of dementing conditions are still very limited and have been disappointing up to now. Prevention studies in AD need to be large, may last for many years and due to that must take into consideration high dropout rates. Enrichment strategies and the development of better screening and measurement tools for asymptomatic or early forms of dementia combined with biomarkers may help to gain more data in the future.

For regulatory purposes, efficacy and safety for symptomatic improvement are typically determined in at least two separate randomized, double-blind, placebo-controlled trials that are each of at least 6 months' duration [8]. As a further requirement, efficacy must be established in a cognitive domain as well as in a functional or global domain. Statistical significant differences must be demonstrated in both primary endpoints. These co-primary clinical outcome measures are currently required for both symptomatic as well as disease-modifying approaches in pivotal phase III studies. However, taking into consideration the natural history of symptom progression in AD and depending on the hypothesized mechanism of action, attempts to establish disease modification are likely to require not only much longer study duration than for symptomatic treatment, but also alternate designs (e.g. staggered start or staggered withdrawal designs). Therefore, clinical trial designs for symptomatic therapy will be different from disease-modifying approaches [8]. Particularly in the latter case the use of biomarkers and the search for adequate surrogate endpoints is encouraged by all stakeholders involved in drug development and harmonization of views is urgently needed [9–13]. Mani [14] outlined that for a claim of disease modification by the FDA, it is required that a therapeutic intervention shows clinical improvement together with modification of the underlying pathophysiological process. Similarly, by European regulators a disease-modifying effect will be considered when the pharmacologic treatment delays the underlying pathological or pathophysiological disease processes and when this is accompanied by an improvement of clinical signs and symptoms of the dementing condition. Consequently, a true disease-modifying effect cannot be established conclusively based on clinical outcome data alone. Such a clinical effect must be accompanied by strong supportive evidence from a biomarker program [8, 15]. Some important thoughts on diagnostic criteria and the role of biomarkers from a regulatory point of view are outlined in the following.

Diagnostic Criteria of Alzheimer's Disease

In our current guidance we follow the criteria for dementia and AD as outlined in the Diagnostic and Statistical Manual of Mental Disorders (DSM-IV-TR of the American Psychiatric Association) or in ICD-10 of the WHO. The probability that a dementia syndrome is caused by AD is essentially based on a history of a steadily progressive course and on the absence of evidence for any other clinically diagnosable cause of the dementia. It can be further specified by using the NINCDS-ADRDA criteria [16]. Almost all clinical trials included patients with probable AD based on these criteria. However, we are highly interested in the ongoing efforts to establish new diagnostic criteria with special emphasis on presymptomatic and early symptomatic stages of AD.

Therefore, a few years ago, mild cognitive impairment (MCI) was proposed as a nosological entity in elderly patients with mild cognitive deficits but without the complete picture of dementia, and as such became an area of high research interest. The rationale behind the development of this term is that an individual patient will pass through a stage of impaired cognition without social or occupational impairment and that the start of treatment in this early stage will result in greater benefits [17, 18]. This new term shows overlapping with other definitions as 'benign senescent forgetfulness', 'age-associated memory impairment', 'age-associated cognitive decline' and 'cognitively impaired not demented'. However, the concept of MCI is still in progress and suffers from several limitations. Estimations of prevalence from epidemiological studies vary greatly depending on the used definitions and criteria. A high proportion of patients diagnosed with MCI returned to normal without progression to dementia, on the other hand in several studies rates of progression from MCI to the full spectrum of dementia up to 12% per year have been described. Preliminary data from clinical trials using cholinesterase inhibitors in patients with MCI have not shown efficacy in the predefined primary endpoints. Thus, MCI has not been considered as a homogeneous clinical entity for regulatory purposes and more work on characterization of meaningful diagnostic criteria is needed, particularly the multiplicity of MCI definitions, the role of etiological subtypes (e.g. amnestic type of MCI) and the development of appropriate assessment tools has to be refined [19, 20].

As members of an international workgroup, Dubois et al. [21, 22] proposed criteria that may help to clarify the MCI controversy. They offered a redefinition of AD as a dual clinico-biological syndrome that can be recognized in vivo, prior to the full onset of dementia, on the basis of (1) a specific core clinical phenotype comprised of an amnestic syndrome of the hippocampal type and (2) supportive evidence from biomarkers of the presence of AD-type biological changes (table 1). AD is diagnosed with the same criteria throughout all symptomatic phases of the disease. The criteria and definitions were further elucidated by the group in 2010 [21]. The major advantage of the Dubois/IWG criteria has been to integrate the use of biomarkers into diagnosis to allow a biologically-based approach to diagnosis independent of disease severity.

Along similar lines, another set of diagnostic criteria has been proposed recently by the National Institute on Aging and Alzheimer Association comprised work groups [23–25]: three clinically relevant diagnoses of preclinical AD, MCI due to AD, and AD dementia with a continuum between and within each stage (NIA-AA criteria). These criteria take into consideration previous clinical definitions of MCI, dementia and AD dementia and incorporate biomarkers electively into the diagnostic workup as research opportunities (table 2). This means that biomarkers are not advocated for routine clinical use for the time being, however in research environments positive biomarkers can be used to stratify patient populations into distinct subgroups with a specific risk profile for AD (up to 17 diagnostic categories).

Both approaches show important similarities and differences and clearly need further validation and standardization work before they can be considered unanimously for pivotal studies in AD. However, from a regulatory point of view, the Dubois/IWG criteria seem to be less complicated and as such more readily applicable in clinical diagnosis and more easily generalizable to everyday clinical practice. As in both approaches biomarkers are key elements for probability of a diagnosis of AD, the regulatory requirements for biomarker use in clinical trials are described.

Table 1. Dubois/IWG research criteria for probable AD: a plus one or more supportive features B, C, D, or E

Core diagnostic criteria	
A	Presence of an early and significant episodic memory impairment that includes the following features: 1 Gradual and progressive change in memory function reported by patients or informants over more than 6 months 2 Objective evidence of significantly impaired episodic memory on testing: this generally consists of recall deficit that does not improve significantly or does not normalize with cueing or recognition testing and after effective encoding of information has been previously controlled 3 The episodic memory impairment can be isolated or associated with other cognitive changes at the onset of AD or as AD advances
Supportive features	
B	Presence of medial temporal lobe atrophy: volume loss in the hippocampus, entorhinal cortex, or amygdalae on MRI with qualitative ratings by visual scoring (referenced to well-characterized population with age norms)
C	Abnormal CSF biomarker: low amyloid β1–42 concentrations, increased total tau concentrations, or increased phospho-tau concentrations, or combinations of the three; or abnormalities in other well-validated markers that will be discovered in the future
D	Specific pattern of reduced glucose metabolism in bilateral temporal parietal regions of functional neuroimaging with PET or with other well-validated ligands, such as Pittsburgh Compound B or FDDNP (2-(1(6-[(2-[^{18}F]fluoroethyl)(methyl)amino]-2-naphthyl)ethylidene) malononitrile)
E	Proven AD autosomal dominant mutation within the immediate family

Table 2. Biomarkers of pathophysiology in the revised NIA-AA diagnostic criteria [23–25]

Biomarkers of brain β-amyloidosis
- Increased uptake on amyloid imaging with PET*
- Decreased CSF amyloid β_{42}*

Biomarkers of neuronal injury (synaptic dysfunction and neuronal degeneration)
- Temporoparietal hypometabolism on ^{18}F-fluorodeoxyglucose PET*
- Medial temporal (hippocampal) atrophy*
- Increased CSF tau/phospho-tau*
- Temporoparietal hypoperfusion on single-photon emission CT

Other less validated biomarkers, biomarkers of collateral damage, or serial biomarkers
- Functional MRI activation studies, resting blood oxygen level-dependent functional connectivity, MRI perfusion, MR spectroscopy, diffusion tensor imaging
- Inflammatory (cytokines) and oxidative stress biomarkers (isoprostanes)
- Rates of brain atrophy

NIA-AA = National Institute on Aging and the Alzheimer's Association.
* Markers included in an early proposal for revised criteria by Dubois et al. [21, 22]

Table 3. Validation of a 'surrogate endpoint'

- Plausible connection between basic science and clinical trials
- Is there a strong, independent, consistent association between surrogate endpoint and clinical outcome (necessary, not sufficient)?
- Evidence from randomized trials that improvements in the surrogate endpoint leads consistently to improvement of the target outcome
- Large, precise, and lasting treatment effects
- Are the likely benefits worth the potential harms and costs?

Regulatory Perspectives on the Use of Biomarkers

The official definition of a biomarker by a NIH working group has been: 'a characteristic that is objectively measured and evaluated as an indicator of normal biological processes, pathogenic processes, or pharmacologic responses to a therapeutic intervention'. A clinical endpoint is a characteristic or variable that measures how a patient feels, functions or survives. A surrogate endpoint is a biomarker that is intended to substitute for a clinical endpoint, which means the change in the surrogate endpoint adequately reflects the change in the clinical endpoint. As such, biomarkers can be used in all stages of drug development to understand the biology and pathophysiology of a disease process, to clarify the mode of action of medicinal products, and to provide information on relevant subpopulations of patients who might respond to treatment or who might be susceptible to side effects. Even the use of a biomarker as primary outcome measure has been used in many drug development programs (phase I and II) after validation of the biomarker [26, 27]. The use of biomarkers as surrogate endpoints in pivotal phase III clinical trials is controversial. The specific biomarker must be validated in the population and clinical setting for which the biomarker is intended – the biomarker can be considered as a 'surrogate endpoint' when it allows substitution for a clinically relevant endpoint that is a direct measure of how a patient feels, functions, or survives and can be expected to predict the effect of the therapy [28–31]. Therefore, a biomarker may have the potential to be a surrogate marker, which then allows the measurement of the effects of a specific treatment (table 3). However, a perfect correlate of the biomarker with disease progression in the untreated state is not sufficient to accept a biomarker as a surrogate for a clinical endpoint and only a few biomarkers are established as surrogate endpoints [32, 33].

Modern neuroimaging, such as various magnetic resonance techniques or amyloid imaging with positron emission tomography are well known examples of such markers. All these markers correlate quantitatively more or less with disease progression. However, as regulators we would look for the link between a treatment-induced change in the biomarker and the desired clinical outcome measure and the link between the treatment-induced change in the biomarker and change of disease

Table 4. Regulatory questions on surrogate markers in AD

- For which clinical outcome the biomarker is used?
- Does the biomarker reliably predict the clinical outcome?
- Does the biomarker reflect effects on pathology and/or pathophysiology for a claim of disease modification?
- Are the effects seen in the biomarkers clinically relevant?
- Allow results seen short-term generalization to long-term?

process. Additionally, there should be high plausibility that based on the assumed mechanism of action of a given medicinal product the disease process will be modified (based on e.g. preclinical models). Medicinal products have been approved on the basis of their effects on validated surrogate markers for many years – examples include antihypertensives, cholesterol-lowering products, or products against glaucoma. However, in AD or other neurodegenerative disorders, none of the imaging or biochemical markers can be considered to be sufficiently validated as a fully developed surrogate endpoint, thereby making their use as primary outcome measures in pivotal efficacy trials unlikely at this time (table 4). Nevertheless, they can be used in ways that allow decision-making on further drug development, they may be used in phase 2 studies (proof-of-concept, dose-finding) as primary prespecified outcome measures or to better define patient populations at risk (enriched populations likely to respond to therapy) for efficacy trials. The use of biomarkers in these ways may permit these studies to be shorter or smaller in size. Ultimately, if any of these biomarkers are found to be acceptable as surrogate endpoints, definitive efficacy trials may also be considerably shorter and/or smaller than studies using more traditional clinical outcomes as primary endpoint [27, 31].

As noted above, however, the reliance on the pharmacological effect on a surrogate marker that has not been validated (that is, for which there has not been shown a strong correlation between the expected change in the surrogate and a beneficial effect of the drug) is fraught with interpretative problems, for example we assume that an efficacious treatment of AD will slow the progression of medial temporal lobe atrophy measured by MRI. However, in the vaccine trial with AN1792 the extent of brain atrophy appeared to increase in patients with increased antibodies and clinical improvement [34]. This was an entirely unexpected outcome and provides evidence that the effects of a treatment (even, potentially, a beneficial treatment) on a chosen surrogate marker can be unpredictable. Of more concern though are those (potential) cases in which the desired effect on the surrogate is achieved. If the clinical effects are unknown, concluding that the drug has a beneficial effect for the patients would be based on the assumption that the desired effect seen on the surrogate will translate into the desired beneficial clinical effect. This assumption may be quite wrong, resulting in the approval of a treatment that has no beneficial (or even, perhaps, a deleterious) effect on the patient. It is for this reason that approval of a treatment for AD based on an effect on a biomarker alone is unlikely at this time. As a general principle, if effects on clinical outcomes can be ascertained in trials of reasonable size and duration, it is likely that approval will not be granted on the basis of studies that examine unvalidated surrogates as sole primary outcomes. Unvalidated surrogates as sole primary outcomes are likely to be considered mostly in those settings in which clinical outcomes cannot be assessed practically, for example in studies examining prevention of AD in which the clinical outcomes (e.g. time of onset of diagnosis of clinical AD) may not occur for many years after treatment initiation.

Conclusion and Consequences for Regulators

Only few medicinal products have been approved for symptomatic improvement in patients with AD. Having in mind the modest effects established in these trials and based on new promising results from basic science and molecular biology in dementia, interest has grown to begin treatment as early as possible and to develop treatment options which will modify the natural course of dementing conditions as AD. This requires better standardization and refinement of diagnostic criteria, and further validation work on the proposed sets of diagnostic criteria is encouraged. Regulators would like to see diagnostic criteria which are easy applicable to define patient populations for clinical trials as well as for use in everyday practice. New diagnostic criteria with emphasis on early disease stages will foster the development of valid and reliable instruments for the assessment of the different domains of AD. Regarding symptomatic improvement focus has changed from global to functional endpoints in addition to the obligatory cognitive endpoint. To establish a disease-modifying effect, several study designs have been proposed with or without incorporation of biomarkers as a potential surrogate endpoint. However, careful and sufficient qualification and validation of the proposed biomarkers is necessary for acceptance as a surrogate endpoint. As this area is so complex with many unresolved development issues, effective and safe medicines will only reach the patient with AD via an early cooperation and continuous sharing of expertise and insight between researchers from basic science, academic medicine, developers from the pharmaceutical industry and regulatory authorities.

Nevertheless, regulatory bodies like FDA and EMA have identified development of biomarkers as high priority in general and particularly in dementia. To foster innovation in this field, consortia like the 'The Biomarker Consortium' [35], 'The Innovative Medicine Initiative' [36] or the ADNI initiative [37] with collaboration of the different stakeholders are highly welcomed and proactively supported. We believe that such public-private partnerships will encourage detection and development of new biomarkers and improve consensus on the requirements of their validation process for biomarkers. Better standardization and validation of the technical aspects of acquisition, measurement, analysis and interpretation of biomarker data and sharing of databases is urgently needed. This will be supported by the ongoing dialogue on harmonization of regulatory practices between EMA and FDA, both of which take into consideration the experience with biomarkers in other fields like oncology or cardiovascular medicine for the field of dementia and AD.

To foster these developments, EMA has established a qualification procedure, which is a new, voluntary, scientific pathway leading to either a CHMP opinion or a scientific advice of novel methodologies on innovative methods or drug development tools. It includes qualification of biomarkers developed by consortia, networks, public/private partnerships, learned societies or the pharmaceutical industry for a specific intended use in drug development. In its first opinion, CHMP addressed the question as to whether the use of two cerebrospinal fluid (CSF)-related biomarkers ($A\beta_{1-42}$ and total tau) are qualified in selecting subjects for trials in early AD as having a high probability of being in the prodromal stage of the disease as defined by the Dubois criteria. Given the relative high sensitivity and moderate specificity, the CSF biomarker signature based on a low $A\beta_{1-42}$ and a high total tau has been considered useful for the enrichment of clinical trial populations [38, 39]. A second qualification opinion for use of low hippocampal volume (atrophy) by MRI in clinical trials in patients with predementia stage of AD has been published for public consultation recently [40]. Further qualification opinions are under way and stakeholders are encouraged to use this possibility [41]. If all stakeholders collaborate, the goal of validation of biomarkers and even surrogate endpoints for use in clinical trials might be reached in order to bring efficacious and safe medicinal products faster to patients with neurodegenerative disorders than they are today.

References

1 Blennow K, Hampel H, Weiner M, Zetterberg H: Cerebrospinal fluid and plasma biomarkers in Alzheimer disease. Nat Rev Neurol 2010;6:131–144.

2 Cummings JL: Optimizing phase II of drug development for disease-modifying compounds. Alzheimers Dement 2008;4:S15–S20.

3 Cummings JL, Doody R, Clark C: Disease-modifying therapies for Alzheimer disease: challenges to early intervention. Neurology 2007;69: 1622–1634.

4 Hampel H, Broich K, Hoessler Y, Pantel J: Biological markers for early detection and pharmacological treatment of Alzheimer's disease. Dialogues Clin Neurosci 2009;11:141–157.

5 Hampel H, Shen Y, Walsh DM, Aisen P, Shaw LM, Zetterberg H, Trojanowski JQ, Blennow K: Biological markers of amyloid β-related mechanisms in Alzheimer's disease. Exp Neurol 2010;223: 334–346.

6 Aisen P: Commentary on 'challenges to demonstrating disease-modifying effects in Alzheimer's disease clinical trials'. Alzheimers Dement 2006;2: 272–274.

7 Cummings J: Challenges to demonstrating disease-modifying effects in Alzheimer's disease clinical trials. Alzheimers Dement 2006;2:263–271.

8 EMA CHMP: Guideline on medicinal products for the treatment of Alzheimer's disease and other dementias. 2008;http://www.ema.europa.eu/docs/en_GB/document_library/Scientific_guideline/2009/09/WC500003562.pdf.

9 Blennow K: Biomarkers in Alzheimer's disease drug development. Nat Med 2010;16:1218–1222.
10 Hampel H, Frank R, Broich K, Teipel SJ, Katz RG, Hardy J, Herholz K, Bokde AL, Jessen F, Hoessler YC, Sanhai WR, Zetterberg H, Woodcock J, Blennow K: Biomarkers for Alzheimer's disease: academic, industry and regulatory perspectives. Nat Rev Drug Discov 2010;9:560–574.
11 Hampel H, Wilcock G, Andrieu S, Aisen P, Blennow K, Broich K, Carrillo M, Fox NC, Frisoni GB, Isaac M, Lovestone S, Nordberg A, Prvulovic D, Sampaio C, Scheltens P, Weiner M, Winblad B, Coley N, Vellas B: Biomarkers for Alzheimer's disease therapeutic trials. Prog Neurobiol 2011;95:579–593.
12 Trojanowski JQ, Vandeerstichele H, Korecka M, Clark CM, Aisen PS, Petersen RC, Blennow K, Soares H, Simon A, Lewczuk P, Dean R, Siemers E, Potter WZ, Weiner MW, Jack CR Jr, Jagust W, Toga AW, Lee VM, Shaw LM: Update on the biomarker core of the Alzheimer's disease neuroimaging initiative subjects. Alzheimers Dement 2010;6:230–238.
13 Vellas B, Coley N, Andrieu S: Disease-modifying trials in Alzheimer's disease: perspectives for the future. J Alzheimers Dis 2008;15:289–301.
14 Mani RB: The evaluation of disease-modifying therapies in Alzheimer's disease: a regulatory viewpoint. Stat Med 2004;23:305–314.
15 Broich K: Outcome measures in clinical trials on medicinal products for the treatment of dementia: a European regulatory perspective. Int Psychogeriatr 2007;19:509–524.
16 McKhann G, Drachman D, Folstein M, Katzman R, Price D, Stadlan EM: Clinical diagnosis of Alzheimer's disease: report of the NINCDS-ADRDA work group under the auspices of Department of Health and Human Services Task Force on Alzheimer's Disease. Neurology 1984;34:939–944.
17 Petersen RC, Knopman DS: MCI is a clinically useful concept. Int Psychogeriatr 2006;18:394–402.
18 Petersen RC, Roberts RO, Knopman DS, Boeve BF, Geda YE, Ivnik RJ, Smith GE, Jack CR Jr: Mild cognitive impairment: ten years later. Arch Neurol 2009;66:1447–1455.
19 Ganguli M: Mild cognitive impairment and the seven uses of epidemiology. Alzheimer Dis Assoc Disord 2006;20:S52–S57.
20 Visser PJ, Brodaty H: MCI is not a clinically useful concept. Int Psychogeriatr 2006;18:402–409.
21 Dubois B, Feldman HH, Jacova C, Cummings JL, Dekosky ST, Barberger-Gateau P, Delacourte A, Frisoni G, Fox NC, Galasko D, Gauthier S, Hampel H, Jicha GA, Meguro K, O'Brien J, Pasquier F, Robert P, Rossor M, Salloway S, Sarazin M, de Souza LC, Stern Y, Visser PJ, Scheltens P: Revising the definition of Alzheimer's disease: a new lexicon. Lancet Neurol 2010;9:1118–1127.
22 Dubois B, Feldman HH, Jacova C, Dekosky ST, Barberger-Gateau P, Cummings J, Delacourte A, Galasko D, Gauthier S, Jicha G, Meguro K, O'Brien J, Pasquier F, Robert P, Rossor M, Salloway S, Stern Y, Visser PJ, Scheltens P: Research criteria for the diagnosis of Alzheimer's disease: revising the NINCDS-ADRDA criteria. Lancet Neurol 2007;6: 734–746.
23 Albert MS, DeKosky ST, Dickson D, Dubois B, Feldman HH, Fox NC, Gamst A, Holtzman DM, Jagust WJ, Petersen RC, Snyder PJ, Carrillo MC, Thies B, Phelps CH: The diagnosis of mild cognitive impairment due to Alzheimer's disease: recommendations from the National Institute on Aging–Alzheimer's Association Workgroups on Diagnostic Guidelines for Alzheimer's Disease. Alzheimers Dement 2011;7:270–279.
24 McKhann GM, Knopman DS, Chertkow H, Hyman BT, Jack CR Jr, Kawas CH, Klunk WE, Koroshetz WJ, Manly JJ, Mayeux R, Mohs RC, Morris JC, Rossor MN, Scheltens P, Carrillo MC, Thies B, Weintraub S, Phelps CH: The diagnosis of dementia due to Alzheimer's disease: recommendations from the National Institute on Aging–Alzheimer's Association Workgroups on Diagnostic Guidelines for Alzheimer's Disease. Alzheimers Dement 2011; 7:263–269.
25 Sperling RA, Aisen PS, Beckett LA, Bennett DA, Craft S, Fagan AM, Iwatsubo T, Jack CR Jr, Kaye J, Montine TJ, Park DC, Reiman EM, Rowe CC, Siemers E, Stern Y, Yaffe K, Carrillo MC, Thies B, Morrison-Bogorad M, Wagster MV, Phelps CH: Toward defining the preclinical stages of Alzheimer's disease: recommendations from the National Institute on Aging–Alzheimer's Association Workgroups on Diagnostic Guidelines for Alzheimer's Disease. Alzheimers Dement 2011;7:280–292.
26 Frank R, Hargreaves R: Clinical biomarkers in drug discovery and development. Nat Rev Drug Discov 2003;2:566–580.
27 Hampel H, Shen Y, Walsh DM, Aisen P, Shaw LM, Zetterberg H, Trojanowski JQ, Blennow K: Biological markers of amyloid β-related mechanisms in Alzheimer's disease. Exp Neurol 2010;223: 334–346.
28 Bucher HC, Guyatt GH, Cook DJ, Holbrook A, McAlister FA: Users' guides to the medical literature. XIX. Applying clinical trial results. A. How to use an article measuring the effect of an intervention on surrogate end points. Evidence-based medicine working group. JAMA 1999;282:771–778.
29 Fleming TR: Surrogate endpoints and FDA's accelerated approval process. Health Aff (Millwood) 2005;24:67–78.

30 Gispen-de Wied CC, Kritsidima M, Elferink AJ: The validity of biomarkers as surrogate endpoints in Alzheimer's disease by means of the quantitative surrogate validation level of evidence scheme (QSVLES). J Nutr Health Aging 2009;13:376–387.
31 Katz R: Biomarkers and surrogate markers: an FDA perspective. NeuroRx 2004;1:189–195.
32 Baker SG, Kramer BS: A perfect correlate does not a surrogate make. BMC Med Res Methodol 2003;3: 16.
33 Baker SG, Kramer BS, Prorok PC: Development tracks for cancer prevention markers. Dis Markers 2004;20:97–102.
34 Fox NC, Black RS, Gilman S, Rossor MN, Griffith SG, Jenkins L, Koller M: Effects of Aβ immunization (AN1792) on MRI measures of cerebral volume in Alzheimer disease. Neurology 2005;64:1563–1572.
35 Woodcock J, Woosley R: The FDA critical path initiative and its influence on new drug development. Annu Rev Med 2008;59:1–12.
36 Council of the European Union: Council regulation (EC) No. 73/2008 of 20 December 2007 setting up the joint undertaking for the implementation of the joint technology initiative on innovative medicines. Official Journal of the European Union 2008: L30/38–L30/51.
37 Cummings JL: Integrating ADNI results into Alzheimer's disease drug development programs. Neurobiol Aging 2010;31:1481–1492.
38 EMA CHMP: Qualification opinion of Alzheimer's disease novel methodologies/biomarkers for bms-708163, 2011, http://www.ema.europa.eu/docs/en_GB/document_library/Regulatory_and_procedural_guideline/2011/02/WC500102018.pdf.
39 Isaac M, Vamvakas S, Abadie E, Jonsson B, Gispen C, Pani L: Qualification opinion of novel methodologies in the predementia stage of Alzheimer's disease: Cerebrospinal-fluid-related biomarkers for drugs affecting amyloid burden – regulatory considerations by European Medicines Agency focusing in improving benefit/risk in regulatory trials. Eur Neuropsychopharmacol 2011;21:781–788.
40 EMA CHMP: Qualification opinion of low hippocampal volume (atrophy) by MRI for use in regulatory trials in predementia stage of Alzheimer's disease, 2011, http://www.ema.europa.eu/docs/en_GB/document_library/Regulatory_and_procedural_guideline/2011/10/WC500116264.pdf.
41 Manolis E, Vamvakas S, Isaac M: New pathway for qualification of novel methodologies in the European Medicines Agency. Proteomics Clin Appl 2011;5:248–255.

Karl Broich, MD
Federal Institute for Drugs and Medical Devices (BfArM)
Kurt-Georg-Kiesinger-Allee 3
DE–53125 Bonn (Germany)
Tel. +49 228 99 307 3255, E-Mail karl.broich@bfarm.de

Hampel H, Carrillo MC (eds): Alzheimer's Disease – Modernizing Concept, Biological Diagnosis and Therapy.
Adv Biol Psychiatry. Basel, Karger, 2012, vol 28, pp 179–188

Perspectives on Alzheimer's Disease: Past, Present and Future

Zaven S. Khachaturian

The Campaign to Prevent Alzheimer's Disease by 2020 (PAD2020), Potomac, Md., USA

Abstract

Alzheimer's disease, as a chronic brain disorder, is the prototype problem for the *Grand Global Challenge* in healthcare. The key quandary is how to balance the relative *'costs'* of investing in research on *'prevention'* (to delay disabilities) with the *scale-price* of healthcare services for burgeoning populations with ever-increasing lifespan. The public policy options are limited to: (a) either invest massive funds into research on *prevention* or, (b) develop plans to *ration* healthcare. The scale of the pending health-economic crisis mandates bold scientific initiative(s) to address this grand global healthcare dilemma. Thus the essential global scientific challenge is to resolve the question of: *'How to accelerate the discovery-development of cures for chronic brain diseases – such as dementia/Alzheimer's disease?'* In spite of remarkable recent advances in the neurobiology of neurodegeneration, there is a growing cynicism regarding current paradigms of drug development due to the (a) lack of effective treatments for dementia, (b) bleak prospects for a dramatic breakthrough in therapy development anytime soon, and (c) inadequate conceptual models of neurodegenerative diseases. Notwithstanding these concerns, many believe the prospect of delaying the onset of disabling symptoms within a decade is an attainable goal, *provided we can surmount several scientific, administrative, and financial impediments*. Among these obstacles the limitations of current conceptual models about etiologies of the disease is an important factor. A quantum shift in current approaches to therapy development requires the adoption of alternative paradigms; such as a *systems failure* model of dementia – based on *general systems theory*.

'The human pain and financial burden of Alzheimer's is so great and the potential breakthroughs in science are so encouraging that a 'Manhattan Project', 'Apollo Project', or 'Human Genome Project' approach to ending Alzheimer's is more than justified. The Alzheimer's Solutions Project is in the best American tradition of solving a big problem with a big vision and a big effort. A public-private partnership is the best collaborative approach to achieve that vision as rapidly as possible. It is the combination of, first, the scale of the crisis and, second, the breadth of the new science which makes this focused, intense investment and project management approach worth implementing.'

Former Speaker Newt Gingrich, Co-chair of the Alzheimer Study Group (ASG), in his testimony to 111th Congress, March 2009.

Quandary of Alzheimer's Disease: Prototype Problem for a Grand Global Challenge

A metaphor for the present-day dilemma facing the global healthcare enterprise was described 3,000 years ago in Greek mythology. The legend of Tithonus depicts a man who cheated death, but whose immortality became a curse rather than a gift. Eos, the goddess of dawn, loved Tithonus so much that she asked Zeus to make him immortal, but she neglected to ask for eternal youth (health) as well. Consequently, Tithonus lived forever but grew extremely old and progressively becoming decrepit and demented.

Modern society is confronting the same quandary as Eos, except on a larger and more complex scale as we confront the economics of healthcare. Now, the *grand global challenge* is to balance the relative *'costs'* of discovering and developing interventions to prevent disability (i.e., the *quest for eternal health*) with the *scale* and *price* of *virtual immortality* (i.e., healthcare for growing populations with an ever-lengthening lifespan).

The public policy options – solutions to this predicament – are limited to two choices: either (1) we invest massive funds to expand research on *prevention* (i.e., disease-modifying interventions to *reduce* the prevalence of costly disabilities), or (2) we start developing plans, the political resolve and the moral fortitude to *ration* healthcare for an aging population.

The Longevity Revolution – Pending Calamity in Healthcare-Economic

Since the Industrial Revolution, improvements in public health, medicine, and nutrition have steadily increased life expectancy. In developed countries, the median age in 1900 was 45 years, today that norm is gradually extending beyond 80 years. This ongoing *'Longevity Revolution'* has dramatically increased the proportion of the population that will survive beyond the 8th, 9th, and 10th decades of life. The *'oldest old'*, those over age 80, comprise the fastest growing segment of the population. A significant proportion, perhaps as high as 4 out of 5, of the population born after the WWII era, the so-called 'baby boomers', are destined to live an additional 30–40 years beyond the traditional retirement age of 62–65 years old.

One of the unintended consequences of this longevity revolution is the unprecedented growth in the demand for health-related programs, services, and products. The pending expansion of the 'health market' is due to the nearly exponential increases in the incidence of disabilities after the age of 65 years. These demographic trends – increasing lifespan and the changing patterns in the prevalence of chronic disorders, such as dementia – foretell a major global public health crisis.

The longevity revolution, which already has had a profound impact on society, will require tectonic shifts in thinking about societal priorities. In light of this pending healthcare tsunami, all developed countries need to reevaluate their current social values systems, paradigms, and public policies across vast arenas of society including: the politics of entitlements, economics of pension plans and social security, labor and retirement, housing and social services, healthcare (including medical insurance, long-term care, and medications) and, most importantly, *investments in research on prevention*.

Simply stated, the scale of the predicament requires new thinking beyond attempts to modify the current systems for financing healthcare and/or delivering services. Traditional constructs to public health will not be sufficient to address the pending healthcare challenges associated with the aging of millions of baby boomers. Unfortunately the core problem of *healthcare in an aging society*, which includes several complex components, does not offer easy or simple solutions for public policies. Thus, a solution might be more readily attained by first focusing on a smaller or more manageable prototype of the larger problem as a preliminary step.

Alzheimer's Disease: A Model for Solving Healthcare Challenges

Among the multitude of contributors to global public health crisis, Alzheimer's disease represents an ideal prototype to serve as a proxy for a number of chronic conditions that require prolonged healthcare and consume costly resources. Neurodegenerative diseases such as Alzheimer's disease and other chronic brain disorders represent a unique class of disabilities not only due to their profound economic impact but also their psychosocial ramifications. The most common clinical features of these unremitting brain conditions – progressive functional impairments of cognition, motor skills, and affect – eventually lead to total dependence on labor-intense care to sustain life. Due to increasing lifespan, the average period of disability for these chronic conditions is gradually being prolonged. At-risk individuals destined to survive beyond the 9th or 10th decade of life now face the prospects of nearly 30–40 years of disability associated with total dependence for personal care, increasing economic burden and deteriorating quality of life.

The enormous scale of the pending crisis in health-economics mandates a bold vision and a compelling scientific agenda to address this grand global healthcare dilemma. To this end, the most critical public policy issues revolve around the question of '*What needs to be done to accelerate the discovery-development of cures for chronic brain diseases – such as dementia/Alzheimer's disease?*'

In short, the grand global challenge – *problem [P]* – can be conceptualized as the product three variables, namely: (1) *increasing numbers [N]* of individuals at risk for various chronic disabilities, (2) *ever-increasing duration [D]* of disabilities, e.g. 30–40 years and, and (3) the *rising cost [C]* of labor-intensive long-term care. Thus, the most effective long-term solution of the problem $[P = N \times D \times C]$ will require formulating public policies and/or initiatives designed to expand global investments in research and development (R&D) programs/initiatives aimed at decreasing the value of *[P]*, either by: (1) *prevention* – reducing the number of people with disability or at risk; (2) *more effective interventions* – shortening the duration of disability, or (3) *lowering the cost of care* – new models of care.

Grand Global Challenge – Prevent Alzheimer by 2020

The primary argument of this paper is that the most effective long-term solution to the looming healthcare crisis is to substantially accelerate the discovery and development of therapies for prevention. A broad spectrum of interventions is needed to maintain the independent functioning of older people and delay disabilities for as long as possible. However, huge investments of funds for research by governments or private entities are not likely to materialize without a compelling scientific agenda.

The justification for a huge infusion of public funds into 'big science' will require not only a credible scientific rationalization but also a realistic strategic plan for effective utilization of sustained investments in research over long periods of time, e.g. 10 years or more. This 10-year strategic business plan needs to defend the heavy investment of funds into global research and development efforts as well as demonstrate the capabilities for streamlined project management approaches, i.e. the adoption of flexible administrative systems that enable rapid decision-making and can handle unexpected opportunities or breakthroughs. Organizational structures and decision-making processes for supporting research projects within existing government agencies, industry, foundations or academia simply cannot meet the needs of the rapidly-evolving scientific world.

In order to mobilize the scientific community towards the objective of formulating such a 10-year business plan, the *Campaign to Prevent Alzheimer's Disease by 2020 (PAD2020)* was launched in 2009 [1]. The mission of the Campaign is to develop a comprehensive action plan for the: (a) expansion of global research capabilities, resources, and infrastructure, and (b) discovery and development of a broad spectrum of interventions targeted towards disease modification and/or prevention of neurodegeneration.

The overall goal of the *PAD2020 Campaign* is to reduce the prevalence of Alzheimer's disease and other brain disorders that affect memory, movement, and mood by 50% within the decade – eventually aiming to prevent the disease entirely. The initiative is based on the premise that a modest delay of 5 years in the onset of disability will reduce the cost and prevalence of the disease by half.

The designation of prevention as target for a global initiative does not imply a promise or a guarantee for disease eradication, but rather, the acceptance of a global goal to mobilize coordinated efforts and a commitment to focus resources towards the achievement of this goal. The enterprise will neither seek nor ask for an assurance of success by the scientific community within the decade; it will merely provide a framework for strategic planning and encourage stretch goals from researchers.

The concept of prevention (defined broadly to include primary, secondary and tertiary prevention) is a clear statement regarding concerted efforts towards a strategic objective to solve the macroeconomic problems of an aging society. The adoption of this goal will not only provide a unifying theme for planning but also offers a conceptual framework for addressing complex relationships among issues concerning science, technology, economics, finance, and public policy.

A global strategic goal to prevent Alzheimer's disease within a decade will be difficult and costly. However, the challenges for this type of a 'big-science' enterprise are no less daunting than other great human endeavors of the past such as the Transcontinental Railroad (1862–1860), championed by Abraham Lincoln and completed in 7 years; the Panama Canal (1904–1914), championed by Theodore Roosevelt and completed in 10 years; the Manhattan Project (1939–1945), championed by Franklin D. Roosevelt and completed in 6 years; the Apollo Program (1961–1969), championed by John F. Kennedy and completed in 8 years, and the Human Genome Project (1990–2000), championed by William J. Clinton and completed in 10 years.

Prevention: A Strategic Goal for Healthcare Crisis

During the last three decades, Alzheimer's research has made remarkable progress in understanding the neurobiology of the disease, and the field has attracted some of the best minds in the world to solve the puzzle of this brain disorder. However, in spite of the important advances, there is a growing disappointment among all stakeholders with the: (a) lack of effective treatments for Alzheimer's disease/dementia; (b) bleak prospects for a dramatic breakthrough in therapy development anytime soon, and (c) inadequate conceptual models of neurodegenerative diseases.

The public's impatience with the slow pace of progress in developing therapies for dementia is understandable. The burden is now on the scientific community to respond to the growing disillusionment by reversing this lack of progress. Notwithstanding the growing cynicism regarding current paradigms of drug development, a significant portion of the scientific community believes that the prospect of delaying the onset of symptoms and eventually preventing Alzheimer's disease within a decade is an attainable goal, *provided we can surmount several scientific, administrative, and financial impediments.*

The full spectrum of efforts in therapy development – from discovery to translation of knowledge on the neurobiology of the disease into practical applications – faces an array of barriers. The critical rate-limiting factors that influence the pace of therapy development include: inadequate funds, high cost of clinical trials/studies, lack of appropriate models and modeling systems, and antiquated management of discovery and development programs. In order to develop strategies to surmount these numerous hurdles, a series of 'Think-Tank' style research planning meetings were organized under the umbrella of Leon Thal Symposia – LTS'07, LTS'08, LTS'09 and LTS'10 [2–6]. Now under the aegis PAD2020 Campaign, these Think-Tank meetings are continuing to seek solutions to the grand global challenge of prevention. PAD2020 Workgroups (WG) have been organized to address essential questions in five generic areas of drug discovery and development for prevention:

- *Science* – What are the scientific and technological obstacles or problems that must be surmounted?

- *Infrastructure/Resources* – What types of infrastructure and resources will be needed by such an undertaking?
- *Regulations/Administrative* – What are the regulatory and/or intellectual property issues that need to be addressed to accelerate therapy development?
- *Organization* – What type of project management team and organizational/ administrative structure will be needed outside of governmental bodies?
- *Finance* – What new and different paradigms for financing are needed for such a massive 10-year international initiative?

Exploring New Conceptual Models of Dementia

Current conceptual models of Alzheimer's disease, which have enjoyed nearly universal acceptance, have proven to be inadequate in providing effective targets for treating complex brain diseases, such as Alzheimer's syndrome, which essential reflects 'systems failure(s)' in an array of interrelated neural networks. The recent disappointments of Dimebon, Semagacestat, and Flurizan, along with the lackluster preliminary results from other ongoing clinical studies, have highlighted the limitations of current conceptual models of the disease, and raised concerns that therapeutic targets currently under investigation may not yield the robust treatment effects (vis-à-vis clinical outcomes) that were anticipated. This uncertainty has begun to recalibrate the thinking in the field about alternative options. Scientists in academia and industry are now ready to reassess many firmly held ideas, assumptions and paradigms of therapy development.

One explanation for the lackluster performance of current drug development paradigms may be the failure of prevailing ideas about the pathogenesis of the disease to fully explain the complex relationship between the clinical and biological phenotypes of the disease. The growing doubts and questions about the ultimate success of hitherto promising therapeutic targets provides a compelling reason for exploring the prospects for different or novel conceptual model(s) of dementia [7].

The enormous scale of the grand global challenge mandates bold, even radical, shifts in thinking about research priorities and paradigms for therapy development. There is an urgent need for new ideas to justify the significant expansion of resources that will be needed to develop a broad spectrum of disease-modifying interventions for prevention and/or the delay of symptom onset.

The PAD2020 virtual *WG on Novel Conceptual Models of Dementia* is an illustrative example of a multinational collaborative effort to promote new thinking about Alzheimer's disease/dementia and other neurodegenerative disorders. The WG will function as an open forum for integrating diverse perspectives about dementia, including knowledge derived from different disciplines and levels of abstractions ranging from molecules to behavior. The endeavor is not intended to be an exercise in denigrating or promoting any particular theories but rather will focus on combining intellectual resources to yield a novel synthesis. The membership of the WG, which consists of key opinion leaders from academia, industry, government and private foundations, will reflect the critical mix of expertise from all relevant areas of science essential for this model building exercise.

Among the array of scientific obstacles to therapy development, the most crucial factors are the limitations of current conceptual models and ideas about etiologies of the disease. The mission of the WG is to lead a comprehensive assessment of all theories on the pathogenesis of Alzheimer's syndrome and proposed alternative conceptual model(s). The aim is to facilitate an objective analysis of the utility of current assumptions concerning the critical cascade of biological events during the full clinical course of the disorder.

The final end product of this endeavor will integrate current knowledge about the syndrome in the formulation of one or more alternative conceptual model(s). The expectation is that this process will identify gaps in knowledge and potential barriers to advancing the discovery and development of disease-modifying or preventive therapies. These new models should: (a) readily account for both *biological* and *clinical* phenotypes of the syndrome; (b) accommodate the full spectrum of the clinical features of the disorder (ranging from preclinical to terminal stages); (c) address the issue of mixed pathologies in Alzheimer's syndrome, including how these pathologies interact and contribute to symptom development, and (d) generate plans for crucial studies needed to determine the validity of these alternative models.

After three decades of remarkable progress of research on the biology of Alzheimer's disease it is now time to take stock of the advances and chart new directions for exploration. Currently there is growing recognition that the cascade of neurobiological processes associated with the disorder start many years before the appearance of any clinical indicators or objectively measurable impairments of function. Thus an additional challenge in formulating a new conceptual model of the syndrome is to enable seamless differentiation, in the preclinical stages of the syndrome, of affected people from those who are unaffected but at elevated risk for dementia. The recent exercise in revising the 27-year-old diagnostic criteria, '*NIA-AA Workgroups on Diagnostic Guidelines for Alzheimer's Disease*' [8–11], has underscored the need for engaging the leaders of the field in a comprehensive evaluation of all prevailing ideas concerning the pathogenesis of Alzheimer's syndrome [1].

'Systems Failure' – A Different Paradigm for Alzheimer's Disease

One illustrative example for a quantum shift in thinking about Alzheimer's syndrome is the proposal to replace current approaches with a *systems failure* model for explaining the neurobiology of this complex brain disorder. The core premise in a systems approach is that the syndrome can be conceptualized in terms of progressive failures in an array of interconnected complex systems or neural network(s). The key explanatory concept of this model is that the syndrome is not the linear result of a unitary etiologic factor.

The operational framework for such a model does not rely on a single theory. Rather, this embedded-set model approach requires understanding the complex interactions among several predisposing biologic events, including changes that occur in sequence and/or in parallel. The model would explain functional relationships among key components of a system and identify approaches to optimize the functioning of the overall system by examining the character of its constitutive components. The explicit challenge for this conceptual framework is to shift the focus of work towards solving the complex interactions necessary to maintain or restore the functionality or optimal performance of 'the system', which could be defined by functionality of synapses, a neuronal network, a well-defined anatomical structure, or a higher-level emergent characteristic within the system, such as memory.

A conceptual model for neurodegenerative disorders based on *general systems theory* is optimally suited to bring about rapid changes in the current paradigms of therapy development. The formulation and validation of a model for a multigenic disorder (e.g. Alzheimer's syndrome) based on systems theory will require: (1) identifying all key etiologic components; (2) understanding the sequence interactions of crucial events, and (3) developing multiple strategies aimed at preserving/maintaining the functionality of the system.

Such a multifactorial model of the syndrome (or neurodegenerative processes in general) will require drastic changes regarding research philosophies in therapy development. In particular, the discovery of new therapeutic targets intended for prevention will require the adoption of a different way of thinking about the full spectrum of pathogenesis. For example, it is well known that the most proximal pathological events associated with the clinical features of the disease are synaptic failure, dendrite pruning, and loss of neurons; therefore, new therapeutic targets should focus on protecting against synaptic dysfunction or repair and regeneration of affected neurons. These therapies for prevention are more likely to succeed when applied in the early, preclinical (or asymptomatic) stages of the disease rather than after symptoms appear. Thus, disease-modifying interventions need to be developed so that they can be delivered decades before the first symptoms appear.

The prime justifications for exploring alternative models of dementia are:

- Presently there are no effective treatments. The lackluster productivity of current drug development paradigms mandates the need to explore different conceptual model(s) of dementia. There is an urgent need to enrich the pipeline for therapy by discovering novel promising therapeutic targets.
- Purely competitive research model of separate groups working by themselves simply will not deliver solutions to this complex problem. A coordinated effort in a non-competitive environment is necessary for validating all of the potential therapeutic targets. Such an open process for sharing information across research groups (in academia and industry) will enable us to: (1) quickly rule in or out

particular approaches, (2) avoid redundancy and blind attacks that lead to ambiguous and/or costly clinical trials, and (3) rank targets with the appropriately focused strategy for development.

- The multitudes of neurobiological processes leading to the syndrome appear to start decades before any objectively measurable impairments of function. Thus, a conceptual model of the syndrome is needed that will enable seamless differentiation of affected from unaffected people, as well as asymptomatic people who are at elevated risk for dementia but in the preclinical stages of the syndrome. The goal is to understand the transition from 'normal' to pathological.
- Alternative models of Alzheimer's syndrome pathogenesis are needed to address the issue of mixed pathologies in differential diagnosis, clinical studies, and treatment.

Arguably, the mission of the WG is a very ambitious undertaking laden with complex challenges. Yet, at this critical junction, the field cannot afford to be deterred by the countless difficulties in identifying the many facets of the underlying biology of Alzheimer's disease and related dementias. The search for alternative conceptual models for Alzheimer syndrome will address the specific needs of: (a) differential clinical diagnosis; (b) discovery and development of novel therapeutic targets; (c) discovery and validation of risk factors, and (d) discovery and validation of surrogate markers of disease progression.

References

1 Khachaturian ZK, Khachaturian AS: Editorial – Prevent Alzheimer's disease by 2020: a national strategic goal. Alzheimer Dement 2009;5:81–84.

2 Khachaturian ZK, Petersen RC, Gauthier S, Buckholtz N, Corey-Bloom JP, Evans W, Fillit H, Foster N, Greenberg B, Grundman M, Sano S, Simpkins J, Schneider LS, Weiner MW, Galasko D, Hyman B, Kuller L, Schenk D, Snyder S, Thomas RG, Tuszynski MH, Vellas B, Wurtman RJ, Snyder PJ, Frank RA, Albert MS, Doody R, Ferris S, Kaye J, Koo E, Morrison-Bogorad M, Reisberg B, Salmon DP, Gilman S, Mohs R, Aisen PS, Breitner JCS, Cummings JL, Kawas C, Phelps C, Poirier J, Sabbagh M, Touchon J, Khachaturian AS, Bain LJ (reporter): Meeting Report – A roadmap for the prevention of dementia: the inaugural Leon Thal Symposium. Alzheimer Dement 2008;4:156–163.

3 Khachaturian ZK, Snyder PJ, Doody R, Aisen P, Comer M, Dwyer J, Frank RA, Holzapfel A, Khachaturian AS, Korczyn AD, Roses A, Simpkins JW, Schneider LS, Albert MS, Egge R, Deves A, Ferris S, Greenberg BD, Johnson C, Kukull WA, Poirier J, Schenk D, Thies W, Gauthier S, Gilman S, Bernick C, Cummings JL, Fillit H, Grundman M, Kaye J, Mucke L, Reisberg B, Sano M, Pickeral O, Petersen RC, Mohs RC, Carrillo M, Corey-Bloom JP, Foster NL, Jacobsen S, Lee V, Potter WZ, Sabbagh MN, Salmon D, Trojanowski JQ, Wexler N, Bain LJ (reporter): A roadmap for the prevention of dementia II: Leon Thal Symposium 2008. Alzheimer Dement 2009;5:85–92.

4 Khachaturian ZK, Cami J, Andrieu A, Avilad J, Boada-Rovira M, Breteler MM, Froelich L, Gauthier S, Gomez-Isla T, Khachaturian AS, Kuller LH, Larson EB, Lopez OL, Martinez-Lage JM, Petersen RC, Schellenberg JD, Sunyer L, Vellas B, Bain LJ (reporter): Meeting Report – Creating a transatlantic research enterprise for preventing Alzheimer's disease. Alzheimer Dement 2009;5:361–366.

5 Khachaturian ZS, Barnes D, Einstein R, Johnson S, Lee V, Roses A, Sager MA, Shankle WR, Snyder PJ, Petersen RC, Schellenberg G, Trojanowski J, Aisen P, Albert MS, Breitner JCS, Buckholtz N, Carrillo M, Ferris S, Greenberg BD, Grundman M, Khachaturian AS, Kuller LH, Lopez OL, Maruff P, Mohs RC, Morrison-Bogorad M, Phelps C, Reiman E, Sabbagh M, Sano M, Schneider LS, Siemers E, Tariot P, Touchon J, Vellas B Bain LJ (reporter): Developing a national strategy to prevent dementia: Leon Thal Symposium 2009. Alzheimer Dement 2010;6:89–97.

6 Khachaturian ZS, Petersen RC, Snyder PJ, Khachaturian AS, Aisen P, de Leon M, Greenberg BD, Kukull W, Maruff P, Sperling RA, Stern Y, Touchon J, Vellas B, Andrieu S, Weiner MW, Carrillo MC, Bain LJ (reporter): Developing a global strategy to prevent Alzheimer's disease: Leon Thal Symposium 2010. Alzheimer Dement 2011;7:127–132.

7 Khachaturian ZK: Editorial – Revised criteria for diagnosis of Alzheimer's disease: National Institute on Aging-Alzheimer's Association diagnostic guidelines for Alzheimer's disease. Alzheimer Dement 2011;7:253–256.

8 Jack CR Jr, Albert MS, Knopman DS, McKhann GM, Sperling RA, Carrillo MC, Thies W, Phelps CH: Introduction to the recommendations from the National Institute on Aging-Alzheimer's Association workgroups on diagnostic guidelines for Alzheimer's disease. Alzheimer Dement 2011;7:257–262.

9 McKhann GM, Knopman DS, Chertkow H, Hyman BT, Jack CR Jr, Kawas CH, Klunk WE, Koroshetz WJ, Manly JJ, Mayeux R, Mohs RC, Morris JC, Rossor MN, Scheltens P, Carrillo MC, Thies B, Weintraub S, Phelps CH: The diagnosis of dementia due to Alzheimer's disease: recommendations from the National Institute on Aging-Alzheimer's Association workgroups on diagnostic guidelines for Alzheimer's disease. Alzheimer Dement 2011;7: 263–269.

10 Albert MS, DeKosky ST, Dickson D, Dubois B, Feldman HH, Fox NC, Gamst A, Holtzman DM, Jagust WJ, Petersen RC, Snyder PJ, Carrillo MC, Thies W, Phelps CH: The diagnosis of mild cognitive impairment due to Alzheimer's disease: recommendations from the National Institute on Aging-Alzheimer's Association workgroups on diagnostic guidelines for Alzheimer's disease. Alzheimer Dementia 2011;7:270–279.

11 Sperling RA, Aisen PS, Beckett LA, Bennett DA, Craft S, Fagan AM, Iwatsubo T, Jack CR Jr, Kaye J, Montine TJ, Park DC, Reiman EM, Rowe CC, Siemers E, Stern Y, Yaffe K, Carrillo MC, Thies B, Morrison-Bogorad M, Wagster MV, Phelps CH: Toward defining the preclinical stages of Alzheimer's disease: recommendations from the National Institute on Aging-Alzheimer's Association workgroups on diagnostic guidelines for Alzheimer's disease. Alzheimer Dement 2011;7:280–292.

Zaven S. Khachaturian, PhD
The Campaign to Prevent Alzheimer's Disease by 2020 (PAD2020)
8912 Copenhaver Drive
Potomac, MD 20854 (USA)
Tel. +1 301 602 4504, E-Mail zaven@pad2020.org

Author Index

Subject Index